Creative
Mathematics

Library of Congress Catalog Card Number 2008938215
ISBN 978-0-88385-750-2
Printed in the United States of America
Current Printing (last digit):
10 9 8 7 6 5 4 3 2 1

Creative Mathematics

H. S. Wall

This book is not a compendium of mathematical
facts and inventions to be read over as a
connoisseur of art looks over paintings in a gallery.
It is, instead, a sketchbook in which readers may
try their hands at mathematical discovery.

The Mathematical Association of America

CLASSROOM RESOURCE MATERIALS

Classroom Resource Materials is intended to provide supplementary classroom material for students—laboratory exercises, projects, historical information, textbooks with unusual approaches for presenting mathematical ideas, career information, etc.

101 Careers in Mathematics, 2nd edition edited by Andrew Sterrett

Archimedes: What Did He Do Besides Cry Eureka?, Sherman Stein

Calculus Mysteries and Thrillers, R. Grant Woods

Conjecture and Proof, Miklós Laczkovich

Creative Mathematics, H. S. Wall

Environmental Mathematics in the Classroom, edited by B. A. Fusaro and P. C. Kenschaft

Exploratory Examples for Real Analysis, Joanne E. Snow and Kirk E. Weller

Geometry From Africa: Mathematical and Educational Explorations, Paulus Gerdes

Historical Modules for the Teaching and Learning of Mathematics (CD), edited by Victor Katz and Karen Dee Michalowicz

Identification Numbers and Check Digit Schemes, Joseph Kirtland

Interdisciplinary Lively Application Projects, edited by Chris Arney

Inverse Problems: Activities for Undergraduates, Charles W. Groetsch

Laboratory Experiences in Group Theory, Ellen Maycock Parker

Learn from the Masters, Frank Swetz, John Fauvel, Otto Bekken, Bengt Johansson, and Victor Katz

Mathematical Evolutions, edited by Abe Shenitzer and John Stillwell

Math Made Visual: Creating Images for Understanding Mathematics, Claudi Alsina and Roger B. Nelsen

Ordinary Differential Equations: A Brief Eclectic Tour, David A. Sánchez

Oval Track and Other Permutation Puzzles, John O. Kiltinen

A Primer of Abstract Mathematics, Robert B. Ash

Proofs Without Words, Roger B. Nelsen

Proofs Without Words II, Roger B. Nelsen

She Does Math!, edited by Marla Parker

Solve This: Math Activities for Students and Clubs, James S. Tanton

Student Manual for Mathematics for Business Decisions Part 1: Probability and Simulation, David Williamson, Marilou Mendel, Julie Tarr, and Deborah Yoklic

Student Manual for Mathematics for Business Decisions Part 2: Calculus and Optimization, David Williamson, Marilou Mendel, Julie Tarr, and Deborah Yoklic

Teaching Statistics Using Baseball, Jim Albert

Writing Projects for Mathematics Courses: Crushed Clowns, Cars, and Coffee to Go, Annalisa Crannell, Gavin LaRose, Thomas Ratliff, Elyn Rykken

MAA Service Center
P.O. Box 91112
Washington, DC 20090-1112
1-800-331-1MAA   FAX: 1-301-206-9789

Foreword

The University of Texas Press in Austin published the first edition of *Creative Mathematics* in 1963. After H.S. Wall's death in 1971, his son Hubert Richard Wall assigned the book's copyright to us at the Educational Advancement Foundation (EAF), and we produced photographic reprints of it in 2006. At our request the Mathematical Association of America (MAA) agreed to update the book and produce this revised edition published in 2008.

Although elements of H.S. Wall's teaching method are described by the author in the Preface, it may be helpful to situate his point of view in an historical context. This book reflects a method of teaching mathematics that was originally practiced by R.L. Moore at the University of Pennsylvania as early as 1911. It began to flourish in the 1920's after he moved to the University of Texas. Referred to by some as *The Moore Method*, it is based upon the notion that mathematics is a creative art and that students should be given the opportunity to experience it in an active inquiry mode rather than as passive recipients of knowledge.

Having arrived at the University of Texas in 1944, Wall was influenced by R.L. Moore's teaching style, and as he and others at the University (such as F. Burton Jones and Hyman Joseph Ettlinger) began to adapt it to their own teaching, it became known as *The Texas Method*. In some modern day circles, it is also known as *inquiry-based learning* (IBL), although that term cuts a much wider swath than the two other descriptors. While important elements of the method are captured in the Preface to this book, those interested in its historical and educational roots would find the following an excellent reference: *R.L. Moore: Mathematician and Teacher*, by John Parker, published by the Mathematical Association of America in 2005. People who have been inspired by this method in recent years have modified the social context in which learning takes place and have introduced a wide variety of discursive formats to enable students to record and reflect upon their mathematical discoveries.

We turn now to a consideration of our decisions regarding changes in this 2008 edition. While the MAA and the EAF have striven to maintain the authenticity of the author's point of view, after considerable deliberation both groups have decided to deviate from one aspect of the 1963 publication. That is, we have replaced a male biased use of pronouns (primarily found in the 1963 preface, but sporadically in other sections as well) with gender neutral language.

While male biased language (using male pronouns to refer to both genders) was *au courant* in most writing up through the middle 1970's, it has been gradually eliminated from most forms of popular and scholarly writing since that time. To have perpetuated male biased language in a text published in 2008 would have signaled an anachronism for many readers. To some people, gender neutral language may appear to be just a cosmetic change, essentially a face lift. Male biased language, however, can be a powerful mechanism that works in subtle ways to threaten and erode the confidence of women who have felt excluded by its persistent use. This is an especially critical concern since today more than half of undergraduate mathematics majors are women as are many of their instructors. While we acknowledge that by removing the male slant (replacing second person singular with third person plural) we on some occasions dilute what would otherwise be a personal tone, we feel that the sacrifice is well worth the advantage.

In addition, we have expanded the content of the 1963 preface in the present edition. In updating this text, the EAF obtained (from the Archives of American Mathematics at the University of Texas) a preface from an unpublished version that was written by H.S. Wall several years prior to his 1963 publication. While it is very similar in content and style to the 1963 preface, it does convey in greater detail and more vibrant language the basic elements of what we referred to above as the *Texas Method*, and in particular to elements of the courses on which the book is based. Since both versions are so consistent, we have minimized repetition and have consolidated the two prefaces in this updated edition.

This revised edition of *Creative Mathematics* is included in the series entitled CLASSROOM RESOURCE MATERIALS published by the MAA. Other volumes in that series, which are listed in the front matter, should be helpful to many instructors who are interested in incorporating variations of IBL in their classrooms. In addition, there are numerous resources available through EAF that readers might like to view in order to gain a fuller understanding of the use of IBL strategies. One of particular interest from an historical perspective is an interview by R.A. Rosenbaum of R.L. Moore in 1964, copyrighted by the MAA in 1966. It is entitled *Challenge in the Classroom* and provides an audio visual introduction to the teaching strategies as well as to the content of the courses depicted in Wall's book.

Readers who would like to request materials from EAF may go to the website educationaladvancementfoundation.org or write to

The Educational Advancement Foundation
2303 Rio Grande Street
Austin, Texas 78705.

Finally, we are most grateful to the MAA for editing and publishing this updated edition of Professor Wall's book—to Don Albers and Beverly Ruedi for overseeing the project, and to Underwood Dudley for his editorial suggestions. In addition to revising the male biased language and updating the typesetting and graphics for this edition, the MAA tightened or clarified language, corrected punctuation, and coherently labeled the figures. We are also grateful to Robert Eslinger and Stephen Brown, who on behalf of EAF reviewed these changes while conscientiously maintaining Professor Wall's meaning and intent. In addition to creating the consolidated preface, they coordinated the input from the different sources that commented on this revision. In addition to those we have already mentioned, the list includes William S. Mahavier, John Neuberger, Albert Lewis, and Hamilton Beazley who made thoughtful comments and suggestions dealing with specific issues. We especially appreciate their input since we no longer have the advice and wisdom of the author to inform us of his specific desires for modifying and updating the book.

Harry Lucas, Jr.
Chairman
The Educational Advancement Foundation

Preface

This book is intended to lead students to develop their mathematical ability, to learn the art of mathematics, and to create mathematical ideas. This is not a compendium of mathematical facts and inventions to be read over as a connoisseur of art looks over the paintings in a gallery. It is, instead, a sketchbook in which readers may try their hands at mathematical discovery.

The American painter Winslow Homer is said to have declared that painters should not look at the works of others for fear of damaging their own directness of expression. I believe the same is true of the mathematician. The fresher the approach the better—there is less to unlearn and there are fewer bad thinking habits to overcome. In my teaching experience, some of my best students have been among those who entered my classes with the least previous mathematical course work. On the other hand, I have usually found it very difficult, if not impossible, to get any kind of creative effort from a student who has had many poor courses in mathematics. This has been true in some cases even though, as it developed later on, the student had very unusual mathematical ability.

The development of mathematical ability does not occur quickly. There are no short cuts. This book is written for the person who seeks an intellectual challenge and who can find genuine pleasure in spending hours and even weeks in constructing proofs for the theorems of one chapter or even a portion of one chapter. It is a book that may be useful for the formal student, but is intended also for the person who is not in school but wants to study mathematics independently. A person who has worked through this book can be regarded as a good mathematician.

In this book, I have tried to say exactly what I mean according to my best understanding of the English language. There are fine shades of meaning in the language used. The little words are especially important. For example, if a man says, "I have *a* son," it is not to be assumed that he does not have two sons. Thus, in this book, a set that contains ten objects contains one object and may contain twenty objects.

I start in the first chapter with certain axioms—statements that are taken for granted—and try to lead students to derive other statements as *necessary consequences* of the axioms. From the axioms and from these new statements—true in the sense that they are consequences of the axioms—still other and deeper statements may be derived. In this way a structure of ideas is built up. Suppose that a student is unable to supply an argument to establish the truth of some statement upon which further developments depend. Rather than seek help from someone or from some other book, the student should take this unsettled thing for granted temporarily and go on to further developments. With additional experience it may later on be possible to go back and fill in the gap. As long as the question remains unsettled, there is a nice problem to work on which, if the creative spirit of this book has been assimilated, will be regarded by the student not as a frustration but as a challenge!

This book reflects a method of teaching that some of us use at the University of Texas. This method of teaching derives from the notion that mathematics is a creative art and that students should be given the opportunity to develop their ability in this art. Starting with this premise, I have, through constant experimentation, developed a certain way of teaching mathematics. In an effort to interest others in this method, I shall try to describe some of its main features.

I have abandoned the lecture method. That is, I do not state a theorem and then proceed to prove it myself. Instead, I try to get students started creating mathematics for themselves as piano teachers start their students creating music—not by lecturing to them but starting at once to develop coordination of mind and muscle. In mathematics this means training the mind to coordinate the right ideas from a set of axioms and definitions and arriving by logical reasoning at a proof of a theorem.

A proof of a theorem consists of a suitable succession of statements each of which is completely justified. It has been my experience that there will be about as many different proofs of certain theorems as there are students who have proved them in my classes. I would not say that one of these proofs is better than another. Different people think in different ways and all should be encouraged. It is thus that new ideas are born!

I have arranged the subject matter of the calculus in such a way that the fundamental ideas may be gradually introduced to, and sometimes even discovered by, students, and all theorems may be proved on the basis of a system of axioms for the number system. This subject matter includes the development of the elementary functions in such a way that trigonometry, for example, is not a prerequisite. Instead of a succession of "obvious" statements as in the lecture method, there is a smaller number of less obvious statements to be proved by students. The questions to be settled become gradually more involved as students develop their powers.

For many of the theorems, I put no definite time limit within which they are to be settled. Frequently, propositions are taken for granted as axioms (with the expectation that they will be settled later on) and freely used, as occasion may arise, in proving other things.

The notion of "covering ground" according to some schedule is completely discarded. The work may seem to progress extremely slowly, especially at the start. Much attention is given to matters of language and logic. Often entire class periods are taken up with these things. To develop clear thinking, it is necessary to develop the ability to make statements that say exactly what is intended. Also, it is necessary to learn to deny statements. Attention to these matters pays large dividends later on. As the work progresses, I am amazed at the accomplishments of the students. In fact, much more ground is covered than under the lecture method.

Very little emphasis is placed upon examinations. I quote from an address given by Professor W.B. Carver, (Thinking versus manipulation, *American Mathematical Monthly*, vol. 44, No. 6 (1937), pages 359–363). "Examination systems, in spite of all efforts to the contrary, seem to influence our teaching in the direction of formalism rather than insight; because it is easy to test a student's manipulative skill and extremely difficult to test his ability to think. . . . may it not be possible that the only really important objective in our teaching of mathematics is something that we will never be able to measure satisfactorily by any kind of test or examination?" I go so far as to give no examinations whatsoever. Instead of the customary examination at the end of each semester, I present a list of problems or topics from which students are asked to select something for a term paper. This is done about two or three weeks before the end of the semester in order to allow students enough time to accomplish something. I believe it is important to remove the fear of examinations so that students may relax and give their brains a chance!

Mathematics is regarded, not as a body of facts, but as a way of thinking and creating ideas. Even if at a given time, all the useful mathematical facts could be assembled and students taught to use them, a short time later on, new facts would be needed to solve new problems. The basic principle is to teach students to think for themselves and to create their own mathematics to solve problems.

Students are encouraged from the outset to develop their own ideas in their own way. If a person's mind works in a certain way effectively, why should a teacher try to change it and perhaps destroy originality? For instance, if a student presents a proof that seems to me to be strange or unnecessarily involved, it rarely occurs to me to point out a different proof. Furthermore, students are proud of their accomplishments in proving a theorem. Should a teacher hurt and discourage them by pointing out some easier (but perhaps not better) proof?

I try to avoid unnecessary names for things and unnecessary symbols. Attaching the names of persons to a theorem might prevent someone from attempting to prove it. For example, imagine a beginner who would not be in awe of "The Bolzano-Weierstrass Theorem." Also, a word or symbol that is a substitute for an idea may very well bury the idea.

Geometrical formulations of definitions and theorems are preferred throughout. Thus in calculus the simple graph rather than the concept of variable is taken as fundamental. By employing geometrical ideas, the sense of sight is utilized and the chance of stirring the imagination is thereby increased. Also the statements in geometric terms can often be simpler than when expressed in other terms.

I point out to students that it is to their advantage not to read proofs in books or even to look into mathematical texts indiscriminately. Some good problem eventually stated in class may thereby be forever spoiled, and originality may be impaired.

I find that this method of teaching can be an inspiration both to students who discover that they have unusual mathematical ability and decide to specialize in mathematics or a mathematical science, and to those of less ability. Students who are unable to prove any but the simplest propositions still get training in language and logical thinking and also obtain the benefit of seeing the presentations by the students who prove their results in class. Moreover, they find out something about what mathematics really is.

This book is the outgrowth of the ideas and inspirations of my many students over the years and I wish to express to them my thanks and my feelings of admiration.

H.S.W.

Contents

Short Biography of H. S. Wall

Creative Mathematics is intimately tied to the author's background and career. Hubert Stanley Wall was born in Rockwell City, Iowa, 2 December 1902. He graduated from the local high school in 1920. He then went to Cornell College in Mt. Vernon, Iowa, where in 1924 he obtained BA and MA degrees with undergraduate work in languages, physics, chemistry and mathematics, and graduate work in mathematics and mathematical physics. Wall's principal influence there—and perhaps the one which led him into mathematics in the first place—was Elmer Moots (1882–1970) to whom in 1944 he dedicated one of his publications. Moots had a reputation at the University of Wisconsin—Madison for referring top-class students to them for graduate work and H.S. Wall was evidently one of these in 1924. He earned his PhD there in 1927 with the dissertation topic: "On the Padé Approximants Associated with the Continued Fraction and Series of Stieltjes."

At Madison his thesis advisor was E. B. Van Vleck, a native of Connecticut who earned his degree in 1893 under Felix Klein, the head of the pre-eminent world center of mathematics at Göttingen University in Germany. When Van Vleck died in 1943 Wall wrote in an article submitted to the *Bulletin of the American Mathematical Society* that he "was profoundly influenced by the teaching and discoveries of Van Vleck," who "loved to explore and survey wide areas, and to teach. In his reading he liked to pick out only the definitions and theorems and then to supply his own proofs."[1]

After receiving his degree Wall, perhaps inspired by Van Vleck, went to Germany himself and did postdoctoral work with Göttingen's most distinguished mathematician, David Hilbert. This stay appears to have lasted for a year or less since he soon began what was to be an active period of research, writing, and publishing—based at Northwestern University, Chicago. Some forty of his fifty-one articles were done while there from 1927 to 1944. In addition he had

---

[1] "The Scientific Work of Edward Burr Van Vleck," H.S. Wall Collection, Archives of American Mathematics, Center for American History, University of Texas at Austin.

five doctoral students and some twenty masters students. From 1937 to 1938 he was a Fellow at the Institute for Advanced Study at Princeton.

In 1939 Wall was instrumental in securing an appointment for Ernst Hellinger at Northwestern, thereby allowing Hellinger, who had been a distinguished professor at Frankfurt University, to be released from Dachau concentration camp and immigrate to the United States. In anticipation of Hellinger's arrival, Wall started studying differential equations from Hellinger's point of view (which was similar to the point of view of Hilbert and Courant). As a result, Wall later wrote *Creative Mathematics* very much in the modern spirit of differential equations in which existence, uniqueness, qualitative study, and numerical computation are emphasized over 'closed form' solutions. Wall worked closely with Hellinger at Northwestern; they published a paper together, and Wall continued to work in the area of what became known as the Hellinger integral.

Their colleague, Walter T. Scott, wrote that Hellinger and Wall

collaborated on an analysis seminar in which they used very brief hints and asked the students to provide proofs of the theorems. This was, let's say, the beginning of the seminar method which he later developed more fully at the University of Texas.... His book, *Creative Mathematics*, was written to try to explain his methods and how he hoped to develop creative talents in his students. [In looking over his letters he wrote to me] his enthusiasm for what his students were doing was amazing. He characterized a student as a potential Hardy or a potential F. Riesz on the basis of his proofs of a theorem or two in a seminar, and this enthusiasm certainly had an effect on his students—there's no doubt.... I remember in talking with him he said that he really felt his students were a much more important contribution to mathematics than was his own research. I think the vote is still being taken—both are important.[2]

In 1944 Wall left Northwestern and spent two years at the Illinois Institute of Technology. Midway through what he anticipated would be a longer stay, he contributed a short observation to the IIT newspaper. Here are excerpts from his writing:

As my first year at the Institute draws to a close, I look forward with a pleasant sense of anticipation to future years here. I feel that there is a spirit of progress which is not satisfied to develop just average

---

[2]Talk given at the presentation awarding the Moore–Wall prize, San Antonio, 1976.

engineers, but which seeks rather to find and develop superior scientists.... Uppermost in the minds of [professors] is the desire to help the student develop the ability to set up and solve problems, and to make free use of mathematics.

I believe progress can be made in that direction. Perhaps the fault lies in the prevalent idea that mathematics is a kit of tools all arranged in little packages. Over the years this has come to be reflected in our textbooks, which in trying to meet the demand for more and more tools in the kit, have reached the point where one can question whether we are teaching mathematics. To illustrate, I believe it would be quite impossible to find out what an integral is from our present calculus text. We teach the integral as a "tool" but fail to teach what the integral is.

I wonder if mathematics isn't rather a state of mind, an analytical mind, which can size up a situation, discard the unimportant, fit disorganized facts into a pattern, and know when a problem is solved. If we approach the teaching of mathematics with this as our axiom, it might be that we could make essential progress. For instance, we would no longer apply the old principle of supplying answers to problems! Part of the scientific mind is the critical ability to know when the problem is solved.

Rather than being a tool subject, I believe mathematics is an art—the purest form of art, in which the mind is the instrument of expression. This is the art which takes chaos and builds from it a magnificent structure of order and reason.

Wall left IIT a year later, in 1946, going to the University of Texas at Austin. In 1948 his book *Analytic Theory of Continued Fractions* appeared and has remained one of the standard works in its field. As he approached the age of fifty there was a shift towards even more attention to cultivating the talents of his students. He devoted himself less to continuing his own impressive research publication record in favor of helping start that of his students.

R.L. Moore had by that time become renowned for the establishment of a method of teaching at Texas that had produced some of the most outstanding mathematicians of the period. Wall found Moore's method, which guided students into gradually more difficult and sophisticated problem solving, developing into independent original research—creating their own proofs and even creating new concepts in the process—very similar to the method of teaching he had been developing. Over the next twenty-five years the two came to share information about their mutual students and to complement each other in the subjects they taught. They formed the core of a small but influential group, with

H.J. Ettlinger being another long-time member, whose teaching styles came to be known collectively as the "Texas method."

Wall died in 1971, having supervised sixty-one doctoral students and some eighty masters candidates in addition.

Albert C. Lewis

Mathematics is a creation of the mind.

To begin with, there is a collection of things that exist only in the mind. Then there is a collection of statements about them, statements that we assume to be true. Starting with these, the mathematician discovers other things, called theorems, and proves them as necessary consequences. This is the pattern of mathematics. The mathematician is an artist whose medium is thought and whose creations are ideas.

1

Numbers

We assume the existence of certain things called *numbers*, some of which are called *counting numbers*, and we take for granted certain statements concerning numbers, called *axioms*. The first few axioms are

Axiom I

If each of x and y is a number, then $x + y$ (read x *plus* y) is a number called the *sum* of x and y. The association with x and y of the sum $x + y$ is called *addition*.

Axiom II

If each of x, y, and z is a number, then $x + (y + z)$ is $(x + y) + z$.

Axiom III

0 is a number such that if x is a number, then $0 + x$ is x.

Axiom IV

If x is a number, then $-x$ is a number such that $x + (-x)$ is 0.

Axiom V

If x and y are numbers, then $x + y$ is $y + x$.

A suitable question may lead to a theorem and one question may lead to another. For example, a study of Axiom III could suggest the question: Is 0 the

1

only number with the property that if x is a number, then $0 + x$ is x? It may be shown on the basis of Axioms III and V that the answer is in the affirmative so that we have

Theorem A. *If $0'$ (read 0 prime) is a number such that if x is a number then $0' + x$ is x, then $0'$ is 0.*

We suggest that the reader try to construct an argument to prove this theorem from the axioms. It can be compared with a sample proof at the end of the chapter.

The number 0—such that if x is a number, then $0 + x$ is x—is called *zero.*

Can Theorem A be improved upon by requiring less of the number $0'$ in order to have as a consequence that $0'$ is 0? Such a result is Theorem A' at the end of the chapter. Before reading further, the student should try to discover and prove some more theorems.

Theorem B. *If each of x and y is a number such that $x + y$ is 0, then y is $-x$.*

If x is a number, *the* number $-x$—such that $x + (-x)$ is 0—is called *minus x* or *the negative* of x. If each of y and x is a number, the number $y + (-x)$ may be denoted by $y - x$. This is called the *difference y minus x* or simply *y minus x*. The association with y and x of the difference $y - x$ is called *subtraction.*

Theorem C. *If each of y and z is a number, the only number x such that $y + x$ is z is the number $z - y$. (That is, $z - y$ is a number x such that $y + x$ is z and if x is a number such that $y + x$ is z, then x is $z - y$).*

Theorem D. *If x is a number, then $-(-x)$ is x.*

Theorem E. -0 *is* 0.

The next few axioms are

Axiom VI

If each of x and y is a number, then $x \cdot y$ (read x *times* y) is a number called the *product* of x and y. The association with x and y of the product $x \cdot y$ is called *multiplication.*

Axiom VII

If each of x, y, and z is a number, then $x \cdot (y \cdot z)$ is $(x \cdot y) \cdot z$.

Axiom VIII
1 is a number such that if x is a number, then $1 \cdot x$ is x.

Axiom IX
If x is a number distinct from 0, then $1/x$ is a number such that $x \cdot (1/x)$ is 1.

Axiom X
If x and y are numbers, then $x \cdot y$ is $y \cdot x$.

Axiom XI
The number 0 is not the number 1.

Axiom XII
If each of x, y, and z is a number, then $x \cdot (y + z)$ is $(x \cdot y) + (x \cdot z)$.

For simplicity, $(x \cdot y) + (x \cdot z)$ may be written without the parentheses, as $x \cdot y + x \cdot z$.

Axioms VI, VII, VIII, IX and X resemble Axioms I, II, III, IV and V, respectively. There is one important difference: the condition *distinct from* 0 in Axiom IX has no counterpart in Axiom IV.

Exercises. Discover and prove theorems concerning multiplication, lettered a, a′, b, c, d, and e that are analogous to theorems A, A′, B, C, D, and E, respectively for addition.

The number 1—such that if x is a number, then $1 \cdot x$ is x—is called *one*. If x is a number distinct from 0, the number $1/x$ of Axiom IX is called the *reciprocal* of x. If z is a number and y is a number distinct from 0, the number x such that $y \cdot x$ is z, namely $z \cdot (1/y)$, is denoted by z/y and called the *quotient* z *over* y. The association with z and y of the quotient z/y is called *division*.

Theorem F. *If x is a number, then $0 \cdot x$ is 0.*

Theorem G. *If each of x and y is a number and $x \cdot y$ is 0, then x is 0 or y is 0* (i.e., if x is not 0, then y is 0).

Theorem H. *If x is a number, then $-1 \cdot x$ is $-x$.*

Theorem J. *If each of x and y is a number, then $(-x) \cdot (-y)$ is $x \cdot y$.*

Note that the meaning of the word *or* is established in Theorem G. The meaning of the word *only* is established in Theorem C.

It is an interesting puzzle to prove Threorem F without the use of Axiom IV, i.e., without the use of negatives of numbers.

The remaining axioms involve the symbol $<$ (read *is less than*). It is convenient to use the symbol $=$ to mean is. Thus, if each of x and y is a number, then $x = y$ means x *is* y; $x < y$ is read x *is less than* y; and the denials may be expressed by $x \neq y$, x *is not* y or x *is distinct from* y; and $x \not< y$, x *is not less than* y.

If each of x and y is a number, the statement $x > y$ means $y < x$ and is read x *is greater than* y or x *exceeds* y. The statement that $x \not> y$ means x *is not greater than* y, i.e., $x = y$ or $x < y$. The statement x *is positive* means $x > 0$ and the statement x *is negative* means $x < 0$.

Axiom XIII
If x and y are numbers, then $x < y$ or $y < x$.

Axiom XIV
If each of x, y, and z is a number and $x < y$, then $x + z < y + z$.

Axiom XV
If each of x, y, and z is a number, and $x < y$ and $0 < z$, then $z \cdot x < z \cdot y$.

Axiom XVI
If each of x, y, and z is a number, and $x < y$ and $y < z$, then $x < z$.

The next axiom tells the distinguishing things about the *counting numbers*, mentioned in the first sentence of this chapter.

Axiom XVII
The number 1 is a counting number. If x is a number such that $x < 1$, then x is not a counting number. If x is a counting number, then $x + 1$ is a counting number and if z is a number such that $x < z$ and $z < x + 1$, then z is not a counting number.

Axiom XVIII
If M is a set each element of which is a number and if there is a number that is less than no number of M, then there is a number k that is less than no number of M that has the property that if k' is a number less than k, then k' is less than *some* number of M. The number k is called *the least number that no number in M exceeds.*

This completes our list of axioms.

Exercise. Show that each of the following statements is a theorem.

(i) If each of x and y is a number, then only one of the following statements is true:

$$x = y, \quad x < y, \quad y < x.$$

(ii) $0 < 1$.

(iii) If x is a number and $0 < x$, then $0 < (1/x)$.

(iv) If x and y are numbers, $0 < x$, and $x < y$, then $(1/y) < (1/x)$.

(v) If x and y are numbers and $y < x$, then there exists a number k such that $0 < k$ and $y + k = x$.

(vi) If x is a number, then $x < x + 1$.

(vii) If each of x, y, u, and v is a number, $x < y$ and $u < v$, then $x + u < y + v$.

(viii) If x and y are numbers and $x < y$, then $-y < -x$.

The statement that the number y is *between* the number x and the number z means that $x < y$ and $y < z$, or $z < y$ and $y < x$. If y is between x and z, then y is between z and x.

Theorem. *Suppose each of x, y, and z is a number. The following statements are equivalent:*

(i) y is between x and z, and

(ii) there exists a number t between 0 and 1 such that $y = t \cdot x + (1 - t) \cdot z$.

Equivalent means that if (i) is true, then (ii) is true, *and* if (ii) is true, then (i) is true. Thus, either statement may be substituted for the other in any argument.

Problems. Suppose each of p and q is a positive number. Find (i) a number that is greater than both p and q and (ii) a positive number which is less than both p and q, i.e., a number which is between 0 and q *and* between 0 and p.

Theorem. *If M is a set each element of which is a number and there is a number which exceeds no number of M, then there is a number h which exceeds no number of M, and if h' is a number greater than h, then h' exceeds some number of M.*

The number h of this theorem is called *the largest number that exceeds no number of M*.

A set each element of which is a number is called a *number set*. A number set containing only one number is a different thing from this number.

The largest number that exceeds no positive number is not in the set of positive numbers.

Theorem. *If M is a number set and each number in M is a counting number, then the largest number that exceeds no number of M belongs to the set M, i.e., M contains a number that is less than any other number of M.*

Exercise. Show that each of the following statements is a theorem.

(i) If x is a number, then there do not exist counting numbers m and n both of which are between x and $x + 1$.

(ii) If h is a number greater than 1 that is not a counting number, then there exists a counting number n such that h is between n and $n + 1$.

(iii) If n is a counting number greater than 1, then $n - 1$ is a counting number.

(iv) If each of m and n is a counting number, then $m + n$ is a counting number and $m \cdot n$ is counting number.

(v) If each of d and k is a positive number, then there exists a counting number n such that $n \cdot d > k$.

Suppose x is a number. The statement that x is a *positive integer* means that x is a counting number; the statement that x is a *negative integer* means that x is the negative of a counting number; the statement that x is an *integer* means that x is a negative integer, x is 0, or x is a positive integer; and the statement that x is a *rational number* means that there exists an integer m and an integer n such that x is the quotient m/n.

Exercise. Show that each of the following statements is a theorem.

(i) If x and y are numbers, there is a rational number between x and y.

(ii) If M is a number set containing 1 such that if x belongs to M then $x + 1$ belongs to M, then every counting number belongs to M. Thus, the set of counting numbers consists of 1, $1 + 1$ or 2, $2 + 1$ or 3, $3 + 1$ or 4, and so forth.

(iii) If each of x and y is a number, then $-x \cdot y = -(x \cdot y)$.

(iv) If each of x and y is a number, then $-(x + y) = -x - y$.

(v) If x is a number, then $x/1 = x$.

(vi) If each of x, y, u, and v is a number with $u \neq 0$ and $v \neq 0$, then

$$\frac{x}{u} \cdot \frac{y}{v} = \frac{x \cdot y}{u \cdot v}.$$

(vii) If x is a number and y is a number distinct from 0, then

$$-\frac{x}{y} = \frac{-x}{y} = \frac{x}{-y}.$$

(viii) If each of a, b, c, and d is a number with $c \neq 0$ and $d \neq 0$, then

$$\frac{a}{c} + \frac{b}{d} = \frac{a \cdot d + b \cdot c}{c \cdot d}.$$

If each of a and b is a number, the product $a \cdot b$ may be written as ab, with the dot omitted. The product aa is denoted by a^2, aa^2 by a^3, aa^3 by a^4, and so forth. The number a^2 is read *a squared*, a^3 is read *a cubed*, a^4 is read *a to the fourth power.*

(ix) If each of a and b is a number, then $(a + b)(a - b) = a^2 - b^2$ and $(a + b)^2 = a^2 + 2ab + b^2$. (Here, $2ab$ means $ab + ab$.)

(x) If each of a and b is a number, then $a^2 + 2ab + b^2 > 0$ unless $a = -b$.

Theorem. *If a is a positive number, there exists a positive number x such that $x^2 = a$.*

A straight line drawn on a writing surface, imagined of indefinite extent, serves as a model for the numbers. Suppose the line is drawn horizontally and extends indefinitely to the right and to the left. A dot on the line is selected and identified with the number 0. Each dot on the line is supposed to be identified with a number and each number with a dot on the line in such a way that, if x and y are numbers and $x < y$, then the dot identified with x is to the left of the dot identified with y, and if X and Y are numbers with $Y - X = y - x$, then the piece of the line with ends identified with x and y is commensurate with the piece of the line with ends identified with X and Y. See Figure 1.1.

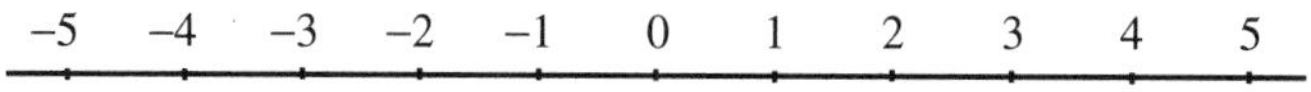

Figure 1.1

We borrow from this model the notions of *interval* and *length* of an interval. The statement that $[a, b]$ is an interval means that a and b are numbers, $a < b$, and $[a, b]$ is the number set to which x belongs only if x is a, x is b, or x is a number between a and b. The length of the interval $[a, b]$ is the positive number $b - a$.

Sample Proofs

Theorem A. *If $0'$ is a number with the property that if x is a number, $0' + x$ is x, then $0'$ is 0.*

Proof. Suppose $0'$ is a number with the property that if x is a number, then $0' + x$ is x. Since 0 is a number, by Axiom III

$$0' + 0 \text{ is } 0$$

and, since $0' + 0$ is $0 + 0'$, by Axiom V

$$0 + 0' \text{ is } 0.$$

But, since $0'$ is a number, by hypothesis, it follows from Axiom III that $0 + 0'$ is $0'$, so that $0'$ is 0, as was to be proved.   □

Theorem C. *If each of y and z is a number, then the only number x such that $y + x$ is z is the number $z - y$.*

Proof. Suppose that each of y and z is a number and that x is a number such that $y + x$ is z. Then

$x + y$	is	z (Axiom V)
$(x + y) + (-y)$	is	$z + (-y)$ (Axioms IV and I)
$x + [y + (-y)]$	is	$z + (-y)$ (Axiom II)
$x + 0$	is	$z + (-y)$ (Axiom IV)
$0 + x$	is	$z + (-y)$ (Axiom V)
x	is	$z + (-y)$ (Axiom III)

i.e., x is $z - y$. Thus, *if* there is a number x such that $y + x$ is z, *then* x is $z - y$.

It remains to be shown that if x is $z - y$, then $y + x$ is z. We have

x	is	$z - y$	(Given)
$y + x$	is	$y + [z - y]$	(Axiom I)
$y + x$	is	$y + [(-y) + z]$	(Axiom V)
$y + x$	is	$[y + (-y)] + z$	(Axiom II)
$y + x$	is	$0 + z$	(Axiom IV)
$y + x$	is	z	(Axiom III)

as was to be proved.   □

Note that *only* x means x *and no other.* Theorem C states that *the* number $z - y$, and *no other* number, has a certain property.

Theorem. *If each of m and n is a counting number, then $m + n$ is a counting number.*

Proof. Suppose m is a counting number. Then by Axiom XVII, $m + 1$ is a counting number. Suppose there is a counting number n such that $m + n$ is *not* a counting number and denote by M the set to which x belongs only if x is a counting number such that $m + x$ is not a counting number. Since 1 is not in M and, by Axiom XVII, no counting number is less than 1, then each number in M is greater than 1 (Axiom XIII). By an earlier theorem, there is a number k in M that is less than any other number of M. Since k is a counting number greater than 1, $k - 1$ is a counting number (by an earlier theorem). Now,

$$m + k = [m + (k - 1)] + 1 \qquad \text{(Axioms II, V, III)}$$

Since $k - 1$ is a counting number, *not in M*, $m + (k - 1)$ is a counting number, and $[m + (k - 1)] + 1$ is therefore a counting number by Axiom XVII, i.e., $m + k$ is a counting number. This contradicts the fact that k is in M. Thus, the set M does not exist—there is no counting number n such that $m + n$ is not a counting number. This establishes the theorem.   □

Theorem A′. *If $0'$ is a number such that for some number x, $0' + x$ is x, then $0'$ is 0.*

2

Ordered Number Pairs

The statement that P is a *point* means that P is an ordered number pair (x, y) having a first number x called the *abscissa* of P and a second number y called the *ordinate* of P. A *point set* is a collection each element of which is a point.

A plane writing surface, e.g., a blackboard, imagined of indefinite extent, may serve as a model for the set of all points. From the set of all horizontal lines on the surface select one and designate it $\underline{0}$ (read 0-*horizontal*) and from the set of all vertical lines on the surface select one and designate it 0 | (read 0-*vertical*). Regard each of these as a model for the set of all numbers, 0 being at the intersection, the positive numbers to the right on $\underline{0}$ and the positive numbers upward on 0 |. If x is a number on $\underline{0}$, denote by x | (read x-*vertical*), the straight line containing x that is vertical, and, if y is a number on 0 |, denote by $\underline{y}$ (read y-*horizontal*), the straight line containing y which is horizontal. The point (x, y) is identified with the intersection of x | and $\underline{y}$ (See Figure 2.1.)

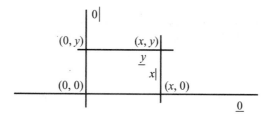

Figure 2.1

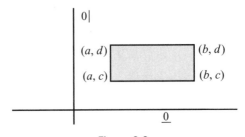

Figure 2.2

In the preceding discussion, *straight line* means a kind of mark on a writing surface—a physical thing. We will now define horizontal and vertical lines as point sets. The statement that $\underline{y}$ is a horizontal line means $\underline{y}$ is the set of all points with ordinate the number y and the statement that $x \mid$ is a vertical line means $x \mid$ is the set of all points with abscissa the number x.

We borrow from the model for the set of all points the notions of *rectangular interval* and *area* of a rectangular interval. The statement that $[ab; cd]$ is a rectangular interval means $[a, b]$ is an interval, $[c, d]$ is an interval, and $[ab; cd]$ is the point set to which (x, y) belongs only if x belongs to $[a, b]$ and y to $[c, d]$. See Figure 2.2.

The area of the rectangular interval $[ab; cd]$ is the positive number $(b - a)(d - c)$.

A point set f such that no two points of f have the same abscissa is called a *simple graph*. For instance, a point set containing only one point is a simple graph and a horizontal line is a simple graph.

The Simple Graph L

We denote by H the simple graph to which the point (x, y) belongs only if x is a positive number and y is the reciprocal of x. Figure 2.3 shows a picture of H. It consists of all the points $(x, (1/x))$ for which $x > 0$.

If P and Q are points of H and P is to the left of Q, then P is *higher* than Q.

In Figure 2.3, x is a number greater than 1 so that the point $(x, 0)$ is to the right of the point $(1, 0)$. The point set S represented by the shaded region is the point set to which the point (u, v) belongs only if u belongs to the interval $[1, x]$ and v is 0, v is $1/u$, or v is a number between 0 and $1/u$.

Question. We have defined the area of the rectangular interval $[ab; cd]$ to be the number $(b - a)(d - c)$. How shall we define a number suitable to be called

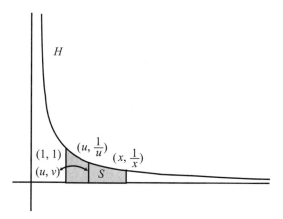

Figure 2.3

the area of the point set S? What would we mean by the area of a field shaped like the shaded region in Figure 2.3?

Rather than try to define area now we will *assume* that there is a number that can be called the area of S. To discuss this number, we introduce the definition: the statement that $[H; a, b]$ is the region determined by H and the interval $[a, b]$ means that $[a, b]$ is an interval of positive numbers and $[H; a, b]$ is the point set to which (x, y) belongs only if x belongs to $[a, b]$ and y to $[0, 1/x]$. In this notation, the shaded region S in Figure 2.3 is $[H; 1, x]$. We assume that the *area* of the region $[H; a, b]$ determined by H and the interval $[a, b]$ is a positive number, denoted by $\int_a^b H$, with the properties

(i)
$$(b - a) \cdot \frac{1}{b} < \int_a^b H < (b - a) \cdot \frac{1}{a}$$

and

(ii) if c is a number between a and b,

$$\int_a^b H = \int_a^c H + \int_c^b H.$$

Condition (i) states that the *area* $\int_a^b H$ of $[H; a, b]$ is a number between the area of the rectangular interval $[ab; 0(1/b)]$ and the area of the rectangular interval $[ab; 0(1/b)]$, the first included in $[H; a, b]$ and the second including

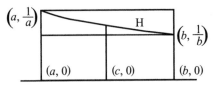

Figure 2.4

$[H; a, b]$. Condition (ii) states that if c is a number between a and b, then the area of $[H; a, b]$ is the sum of the area of $[H; a, c]$ and the area of $[H; c, b]$ (See Figure 2.4.)

If p and q are numbers and $p < q$, the number $(1/2)(p + q)$ is a number r called the *average* of p and q which has the property of being between p and q and also the property that $r - p = q - r$, i.e., the length of $[p, r]$ is the length of $[r, q]$. Thus, the number r *bisects* the interval $[p, q]$. If x is a number between p and q, then x differs from this number r by less than $(1/2)(q - p)$, i.e., by less than one-half the length of $[p, q]$. *It follows that $\int_a^b H$ differs from the average of $(b - a) \cdot (1/b)$ and $(b - a) \cdot (1/a)$ by less than $\frac{(b-a)^2}{2ab}$.* Thus,

$$\int_a^b H \text{ is approximately } \frac{b^2 - a^2}{2ab}, \text{ with error } < \frac{(b - a)^2}{2ab}.$$

If, for example, a is not less than 1 and $b - a = .1$, then the error in the approximation is less than .005, so that the approximate formula is accurate to two decimal places. $A \doteq B$ means A is approximately equal to B.

Example.

$$\int_1^{1.1} H \doteq \frac{(1.1)^2 - 1^2}{2.2} = .095,$$

$$\int_{1.1}^{1.2} H \doteq \frac{(1.2)^2 - (1.1)^2}{(2.2)(1.2)} = .087,$$

Then, by (ii),

$$\int_1^{1.2} H = \int_1^{1.1} H + \int_{1.1}^{1.2} H \doteq .182.$$

These approximations are accurate to about two decimal places.

Exercise. Complete the following table of approximations.

x	$\int_1^x H$
1.1	.095
1.2	.182
1.3	
1.4	
1.5	
1.6	
1.7	
1.8	
1.9	
2.0	

Definition. L denotes the simple graph to which the point (x, y) belongs only if x is a positive number and

$$y = \begin{cases} -\int_x^1 H, & 0 < x < 1, \\ 0, & x = 1, \\ \int_1^x H, & 1 < x. \end{cases}$$

Problems

1. Find any evidence you can to support the following statement: If x is a number greater than 1, then

$$\int_{1/x}^1 H = \int_1^x H.$$

2. Find any evidence you can to support the following statement: If c is a positive number, then the area of $[H; a, b]$ is the area of $[H; ca, cb]$, i.e.,

$$\int_a^b H = \int_{ca}^{cb} H, \quad 0 < a < b \text{ and } 0 < c.$$

3. Use the computations of the last exercise and the statement in Problem 1
to sketch the simple graph L.

Definition. If f is a simple graph, the ordinate of that point of f whose abscissa
is the number x is denoted by $f(x)$ (read f of x).

For example, if x is a positive number, $H(x) = 1/x$. Also, $L(1) = 0$,
$L(1.1) = .095$, $L(1.2) = .182$, the last two are approximations to three decimal places, and if x is a positive number,

$$L(x) = \begin{cases} -\int_x^1 H, & 0 < x < 1, \\ 0, & x = 1, \\ \int_1^x H, & 1 < x. \end{cases}$$

Exercise. Assume the properties of area of $[H; a, b]$ and the statements in
Problems 1 and 2 to prove

(i) $\int_a^b H = L(b) - L(a)$,   if $0 < a < b$,

(ii) $L(b) - L(a) = L(ca) - L(cb)$,   if $0 < a < b$ and $0 < c$,

(iii) if each of x and y is a positive number, then

$$L(xy) = L(x) + L(y),$$
$$L(x/y) = L(x) - L(y),$$
$$\text{and } L(1/x) = -L(x).$$

Use these formulas and computations already made to find approximations
to $L(3), L(5), L(7), L(11), L(13), L(17), L(19)$. For example, knowing approximations to $L(1.5)$, which is $L(3/2)$ or $L(3) - L(2)$, and to $L(2)$, an approximation to $L(3)$ can be obtained. Next, use these results to find approximations
to $L(.1), L(.2), L(.3), L(.4), L(.5), L(.6), L(.7), L(.8)$, and $L(.9)$.

The problem of defining the area of $[H; a, b]$ involves picking out a certain
number from the set of all numbers. The axiom preeminently adapted to this
purpose is Axiom XVIII. First, we try to define a number set M no number
of which exceeds some number k and such that the *least* such number k not
exceeded by any number of M has the properties required of the number
we are trying to define. Geometry could lead to an idea for setting up the
number set M. For the present we shall continue to assume that $[H; a, b]$ has
an area with the properties already specified, and that the statements made in

Problems 1 and 2 are true. We then have the formulas of the last exercise at our disposal.

The Logarithm

If a is a positive number distinct from 1 and x is a positive number, *the logarithm of x to the base a*, denoted by $\log_a x$, is the number $\frac{L(x)}{L(a)}$:

$$\log_a x = \frac{L(x)}{L(a)}.$$

The logarithm to the base 10 is called the *common logarithm*. Thus, $\log_{10} 2 = \frac{L(2)}{L(10)} = .301$. Common logarithms are used to reduce the problems of multiplication and division to the simpler problems of addition and subtraction. Tables have been computed for this purpose. Also, the slider rule furnishes a mechanical means to the same end.*

Exercise. Show that, if each of x and y is a positive number and each of a and b is a positive number distinct from 1, then

$$\log_a xy = \log_a x + \log_a y, \qquad \log_a \frac{x}{y} = \log_a x - \log_a y,$$

and

$$\log_a \frac{1}{x} = -\log_a x.$$

Moreover, $(\log_a b)(\log_b a) = 1$. Show that if n is an integer,

$$\log_{10}(10^n x) = n + \log_{10} x.$$

For instance, $\log_{10} 20 \doteq 1.301$, and $\log_{10} .2 \doteq .301 - 1$. As usual 10^{-1} means .1, 10^{-2} means .01, etc.

The statement that S is an inner sum for the region $[H; a, b]$ means there exists a finite collection G of nonoverlapping intervals filling up $[a, b]$ such that if the length of each interval $[p, q]$ in G is multiplied by $1/q$, then the sum of all the products so formed is S.

An inner sum for $[H; a, b]$ is the sum of the areas of one or more rectangular intervals. The simplest is $(b - a) \cdot (1/b)$. The next simplest is the sum of the areas of two rectangular intervals: $(c - a) \cdot (1/c) + (b - c) \cdot (1/b)$, where

---

*Editor's note: Tables and slide rules were particularly useful before the advent of calculators and computers.

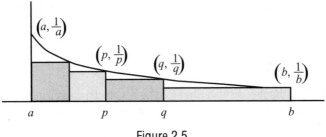

Figure 2.5

$a < c < b$. The inner sum in Figure 2.5 is the sum of the areas of five rectangular intervals.

Every inner sum for $[H; a, b]$ is less than $(b - a) \cdot (1/a)$. Hence, by Axiom XVIII, there is a least number k which no inner sum for $[H; a, b]$ exceeds. This number k is taken as the area of $[H; a, b]$ and denoted by the symbol $\int_a^b H$ (read: the area of the region determined by H and $[a, b]$).

Theorem. *Suppose a and b are positive numbers, $a < b$ and c is a positive number. Then*

(i) $(b - a) \cdot \frac{1}{b} < \int_a^b H < (b - a) \cdot \frac{1}{a}.$

(ii) *If c is between a and b, $\int_b^a H = \int_a^c H + \int_c^b H.$*

(iii) $\int_a^b H = \int_{ca}^{cb} H.$

Establishing this theorem will give substance to this chapter, which up to now has been based on conjecture.

For a proof of (ii), it may be useful to turn to our straight line model for the numbers. If $x = \int_a^c H$, $y = \int_c^b H$, and $z = \int_a^b H$, then only one of the following three statements is true:

$$z < x + y, \quad z > x + y, \quad \text{or} \quad z = x + y.$$

If the first two statements can be proved impossible, then the last is true, i.e., (ii) is true. The first two statements may be pictured as in Figure 2.6:

| 0 | x | y | z | x+y | | 0 | x | y | x+y | z |

Figure 2.6

No inner sum for $[H; a, c]$ exceeds x but, if x' is a number less than x, *some* inner sum for $[H; a, c]$ exceeds x' and a similar statement holds for y and z.

Properties of L

We state here some properties of the simple graph L that may be proved using properties already discovered, e.g.,

$$L(b) - L(a) = \int_a^b H, \qquad 0 < a < b,$$

$$(b - a) \cdot \frac{1}{b} < \int_a^b H < (b - a) \cdot \frac{1}{a},$$

and

$$L(xy) = L(x) + L(y), \qquad x > 0, y > 0.$$

   i. If h is a number, there exists a point of L above the horizontal line $\underline{h}$ and a point of L below the horizontal line $\underline{h}$.

   ii. If x and y are positive numbers and $x < y$, then $L(x) < L(y)$.

   iii. If h is a number, the horizontal line $\underline{h}$ contains a point of L but does not contain two points of L.

Definition. The X-projection of the simple graph f is the number set to which x belongs only if x is the abscissa of a point of f. The Y-projection of the simple graph f is the number set to which y belongs only if y is the ordinate of a point of f.

   The X-projection of L is, by definition of L, the set of all *positive* numbers. From property (iii), *the Y-projection of L is the set of all numbers.*

The Simple Graph E

The point set to which (x, y) belongs only if (y, x) belongs to L is a simple graph that we denote by E. The X-projection of E is the set of all numbers and the Y-projection of E is the set of all positive numbers. Figure 2.7 shows a sketch of L and of E.

   From the definition of E and properties of L, the reader may establish the following properties of E.

   i. If x and y are numbers and $x < y$, then $E(x) < E(y)$.

   ii. If x is a number, $L\{E(x)\} = x$, and if x is a positive number, $E\{L(x)\} = x$.

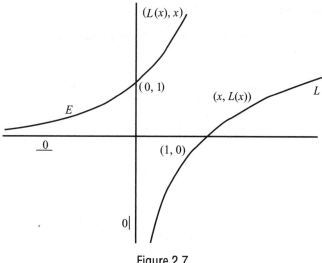

Figure 2.7

iii. If each of x and c is a number, then

$$E(x + C) = E(x) \cdot E(c).$$

iv. If e denotes the number $E(1)$, then $L(e) = 1$, and $2.7 < e < 2.8$.

v. $E(1) = e$, $E(2) = e^2$, $E(3) = e^3$, and if n is a positive integer, $E(n) = e^n$.

vi. If x is a number, $E(-x) = \frac{1}{E(x)}$.

Powers and Roots

Definition. If a is a positive number and x is a number, then a^x (read *a to the x*) is the number $E\{xL(a)\}$:

$$a^x = E\{xL(a)\}.$$

In particular, $e^x = E(x)$.

Examples. If a is a positive number, $a^0 = 1$ and $a^1 = a$ and, if n is a positive integer, $a^n = a \cdot a^{n-1}$. If x is a number, $1^x = x$.

Theorem. *If each of a and b is a positive number, and each of x and y is a number, then*

(i) $a^x a^y = a^{x+y}$,

(ii) $(a^x)^y = a^{xy}$,

(iii) $a^x b^x = (ab)^x$,

(iv) $a^{-x} = \dfrac{1}{(a^x)}$,

(v) $\dfrac{a^x}{a^y} = a^{x-y}$,

(vi) $\dfrac{a^x}{b^x} = \left(\dfrac{a}{b}\right)^x$,

(vii) $L(a^x) = x L(a)$,

(viii) $\log_b a^x = x \cdot \log_b a$,

(ix) *If n is a positive integer,* $(a^{1/n})^n = a$.

Definitions. The number $a^{1/2}$ is called the positive square root of the positive number a and may be denoted by $\sqrt{a}$. This is the only positive number whose square is a. If n is an integer greater than 2, $a^{1/n}$ is called the positive nth root of a and may be denoted by $\sqrt[n]{a}$.

The Simple Graph Q

Q denotes the simple graph whose X-projection is the set of all nonnegative numbers defined by

$$Q(x) = \begin{cases} 0, & x = 0, \\ x^{1/2}, & x > 0. \end{cases}$$

Figure 2.8 shows a sketch of the simple graph Q.

Definition. If x is a number, $|x|$ (read *the absolute value of x*) is the number $Q(x^2)$.

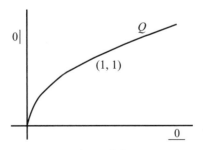

Figure 2.8

Exercise. Show that if each of x and y is a number, then

$$|x + y| \not> |x| + |y|,$$

i.e., $|x + y|$ is less than or equal to $|x| + |y|$. This may be written $|x + y| \leq |x| + |y|$.

3

Slope

Straight Lines

The statement that f is the straight line of *slope* m containing the point (a, b) means that m is a number and f is the simple graph such that for every number x

$$f(x) = m \cdot (x - a) + b.$$

Exercise

(i) Show that if (x, y) and (u, v) are points of the straight line of slope m containing the point (a, b), then

$$m = \frac{y - v}{x - u}.$$

(ii) Sketch the straight line of slope 1 containing the point $(0, 0)$ and the straight line of slope -1 containing the point $(0, 0)$.

(iii) Suppose c is a positive number. Determine a number m such that the straight line of slope m containing the point $(c, L(c))$ has none of its points below L.

(iv) Suppose c is a positive number. Show that the straight line of slope $-(1/c)^2$ containing the point $(c, (1/c))$ has none of its points above H.

(v) Suppose W is the simple graph such that if x is a number, then $W(x) = x^2$. If c is a number, determine a number m such that the straight line of slope m containing (c, c^2) has none of its points above W.

(vi) Suppose P and Q are points of a straight line of slope m. Show that if P is to the left of Q, then P is lower than Q or P is higher than Q according as $m > 0$ or $m < 0$.

Computation Formulas for L

It is possible to conjecture formulas for computation of $L(c)$ from geometrical considerations. We suppose $c > 1$. Since $(c - 1) \cdot (1/c)$, i.e., $(c - 1)/c$, is an inner sum for $[H; 1, c]$, it follows that $L(c) > (c - 1)/c$, so that there exists a positive number k_1 such that

$$L(c) = \frac{c - 1}{c} + k_1, \qquad 0 < k_1.$$

To obtain a number larger than k_1, we see from Figure 3.1 that k_1 appears to be less than the area

$$\frac{1}{2}(c - 1)\left(1 - \frac{1}{c}\right) \quad \text{or} \quad \frac{1}{2}\left(\frac{c - 1}{c}\right)^2 \cdot c$$

of triangle ABC. Thus,

$$L(c) = \frac{c - 1}{c} + k_1, \qquad 0 < k_1 < \frac{1}{2}\left(\frac{c - 1}{c}\right)^2 \cdot c.$$

To obtain a second approximation to $L(c)$, we use (iv) of the last exercise. The straight line f containing the point $(c, (1/c))$ and having slope $-(1/c)^2$,

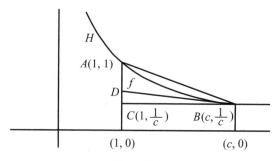

Figure 3.1

defined for every number x by

$$f(x) = -\left(\frac{1}{c}\right)^2 (x - c) + \frac{1}{c},$$

has none of its points above H (See Figure 3.1). Hence $L(c)$ is less than $(c - 1)/c$ plus the area of triangle BCD:

$$L(c) = \frac{c - 1}{c} + \frac{1}{2}\left(\frac{c - 1}{c}\right)^2 + k_2, \qquad 0 < k_2.$$

By analogy with the first approximation formula, we conjecture that $k_2 < \frac{1}{3}(\frac{c-1}{c})^3 \cdot c$; i.e.,

$$L(c) = \frac{c - 1}{c} + \frac{1}{2}\left(\frac{c - 1}{c}\right)^2 + k_2, \qquad 0 < k_2 < \frac{1}{3}\left(\frac{c - 1}{c}\right)^3 \cdot c.$$

Having gone this far, we now conjecture

$$L(c) = \frac{c - 1}{c} + \frac{1}{2}\left(\frac{c - 1}{c}\right)^2 + \frac{1}{3}\left(\frac{c - 1}{c}\right)^3 + k_3, \qquad 0 < k_3 < \frac{1}{4}\left(\frac{c - 1}{c}\right)^4 \cdot c$$

and so on.

If our formulas are correct, they clearly give better approximations the nearer c is to 1. To obtain experimental evidence for our formulas, we note that $L(2)$, $L(3)$, $L(5)$, and $L(7)$ can be found from $L(\frac{25}{24})$, $L(\frac{36}{35})$, $L(\frac{81}{80})$, and $L(\frac{225}{224})$ or by means of other fractions greater than 1 involving only the integers 2, 3, 5, and 7 as factors. In fact,

$$L\left(\frac{25}{24}\right) = -3L(2) - L(3) + 2L(5),$$

$$L\left(\frac{36}{35}\right) = 2L(2) + 2L(3) - L(5) - L(7),$$

$$L\left(\frac{81}{80}\right) = -4L(2) + 4L(3) - L(5),$$

$$L\left(\frac{225}{224}\right) = -5L(2) + 2L(3) + 2L(5) - L(7).$$

If $c = \frac{25}{24}$, so that $\frac{c-1}{c} = \frac{1}{25} = .04$, then we find by means of the fourth formula

$$L\left(\frac{25}{24}\right) = .04 + .0008 + .000021333 + .00000064 + k_4,$$

where $0 < k_4 < .000000021$, so that $L(\frac{25}{24})$ can be expressed as .0408220 correct to 7 decimal places. Likewise

$$L\left(\frac{36}{35}\right) \doteq .0281709, \qquad L\left(\frac{81}{80}\right) \doteq .0124225, \qquad L\left(\frac{225}{224}\right) \doteq .0044543.$$

We then find the values

$$L(2) \doteq .69315, \quad L(3) \doteq 1.09861, \quad L(5) \doteq 1.60944, \quad L(7) \doteq 1.94591.$$

From these values, $\log_{10} 2 = \frac{L(2)}{L(10)} = .30103$, agreeing with the correct value to five decimal places. Also, from our computation formulas we get $L(\frac{10}{9}) = .10536$, and from the above values, $L(\frac{10}{9}) = L(2) + L(5) - 2L(3) \doteq .10536$.

The experimental evidence tends to support our formulas.

Computation of e

The number $E(1)$, denoted by e, is the number x such that $L(x) = 1$. Since $L(2) = .69315$ and $L(3) = 1.09861$, we see that $2 < e < 3$ and e is nearer to 3 than to 2. We have

$$L(2.7) = L\left(\frac{27}{10}\right) = 3L(3) - L(2) - L(5) \doteq .99324,$$

so that $e > 2.7$. Since $L(2.8) = L(\frac{14}{5}) = L(2) + L(7) - L(5) \doteq 1.02962$, we have that $e < 2.8$. Thus

$$2.7 < e < 2.8.$$

Also, e is nearer 2.7 than 2.8. To compute $L(2.71)$, note that $2.71 = \frac{271}{270} \cdot \frac{27}{10}$, so that $L(2.71) = L(\frac{271}{270}) + L(2.7) \doteq .99694$ and so that $e > 2.71$, $L(2.72) = L(\frac{272}{271}) + L(\frac{271}{270}) + L(2.7) \doteq 1.00062$, so that

$$2.71 < e < 2.72.$$

It may be shown that to five decimal places, $e = 2.71828$.

Slope of a Simple Graph

The statement that the point P is *between* the simple graph f and the simple graph g means that if $P = (x, y)$ then the ordinate y of P is between $f(x)$ and $g(x)$.

The statement that the simple graph f *has slope at the point P* means that P is a point of f such that each two vertical lines with P between them have

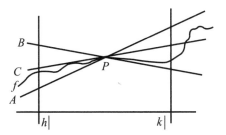

Figure 3.2

between them a point of f distinct from P, and that there exists a number m such that, if A is a straight line of slope greater than m containing P, and if B is a straight line of slope less than m containing P, then there exist vertical lines $h|$ and $k|$ with P between them such that every point of f distinct from P between $h|$ and $k|$ is between A and B.

If the simple graph f has slope at the point P, then there is only one number m having the property described in the definition. This number m is called the *slope* of f at P, and the straight line of slope m, containing P is called the *tangent line* to f at P. See Figure 3.2.

Theorem. *If the simple graph f has slope at the point P, if each two vertical lines with P between them have between them a point of f to the left of P and a point of f to the right of P, and if g is a straight line containing P such that no point of f is above g, then g is the tangent line to f at P.*

Theorem. *Suppose that f is a simple graph, that P is the point $(x, f(x))$ of f such that each two vertical lines with P between them have between them a point of f distinct from P, and that m is a number. The following two statements are equivalent:*

(i) *f has slope m at P, and*

(ii) *if c is a positive number, there exists a positive number d such that, if y is the abscissa of a point of f whose distance from x is less than d, then*

$$f(y) - f(x) = (y - x) \cdot m + (y - x) \cdot \begin{bmatrix} a \text{ number between} \\ -c \text{ and } c \end{bmatrix}.$$

Suggestion. Recall that if equal of u and v is a number, then the number x is *between* u and v only in case there is a number t between 0 and 1 such that $x = (1 - t) \cdot u + t \cdot v$.

Examples.

(i) If each of x and y is a number,

$$y^2 - x^2 = (y - x) \cdot 2x + (y - x) \cdot [y - x].$$

Hence, if c is a positive number and y differs from x by less than c,

$$y^2 - x^2 = (y - x) \cdot 2x + (y - x). \begin{bmatrix} \text{a number between} \\ -c \text{ and } c \end{bmatrix}.$$

Thus, the simple graph W defined for each number x by $W(x) = x^2$ has slope $2x$ at the point (x, x^2).

(ii) If x and y are positive numbers,

$$H(y) - H(x) = (y - x) \cdot \left[-\left(\frac{1}{x}\right)^2 \right] + (y - x) \cdot \left[\frac{x - y}{x^2 y} \right].$$

If c is a positive number and d a positive number less than both $\frac{x}{2}$ and $\frac{cx^3}{2}$, then if y differs from x by less than d

$$H(y) - H(x) = (y - x) \cdot \left[-\left(\frac{1}{x}\right)^2 \right] + (y - x). \begin{bmatrix} \text{a number between} \\ -c \text{ and } c \end{bmatrix}.$$

Thus H has slope $-(1/x)^2$ at the point $(x, H(x))$.

In these examples we had reason to believe that if the graphs had slopes, then $2x$ was the slope of the first and $-(1/x)^2$ that of the second. We have verified that the graphs have these slopes. How should we proceed if we have no previous knowledge as to what the slope should be?

(iii) Consider the simple graph Q. If x and y are positive numbers,

$$Q(y) - Q(x) = \frac{[Q(y) - Q(x)][Q(y) + Q(x)]}{Q(y) + Q(x)} = \frac{y - x}{Q(y) + Q(x)}.$$

Thus, if $U = (x, Q(x))$ and $V = (y, Q(y))$, then the slope of the straight line containing U and V is

$$\frac{Q(y) - Q(x)}{y - x} = \frac{1}{Q(y) + Q(x)}.$$

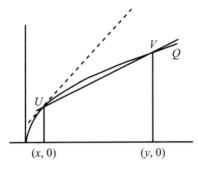

$(x, 0)$ $(y, 0)$

Figure 3.3

(See Figure 3.3.) The nearer y is to x, the nearer this slope is to $\frac{1}{2Q(x)}$, so we conjecture that the slope of Q at the point U is $\frac{1}{2Q(x)}$. We then write

$Q(y) - Q(x)$

$$= (y - x) \cdot \frac{1}{Q(y) + Q(x)}$$

$$= (y - x) \cdot \frac{1}{2Q(x)} + (y - x) \cdot \left[\frac{1}{Q(y) + Q(x)} - \frac{1}{2Q(x)} \right]$$

$$= (y - x) \cdot \frac{1}{2Q(x)} + (y - x) \cdot \frac{Q(x) - Q(y)}{[Q(y) + Q(x)]2Q(x)}$$

$$= (y - x) \cdot \frac{1}{2Q(x)} + (y - x) \cdot \frac{x - y}{[Q(y) + Q(x)][Q(y) + Q(x)]2Q(x)},$$

from which it is easy to prove that our conjecture is correct. In fact, if c is a positive number and $d = c \cdot 2[Q(x)]^3$, then if y is a positive number which differs from x by less than d

$$Q(y) - Q(x) = (y - x) \cdot \frac{1}{2Q(x)} + (y - x) \cdot \left[\begin{array}{c} \text{a number between} \\ -c \text{ and } c \end{array} \right].$$

Definition. If f is a simple graph, $D_x f$ denotes the slope of f at the point of f whose abscissa is x.

Examples.

$$D_x W = 2x,$$

$$D_x H = -\left(\frac{1}{x}\right)^2,$$

$$D_x Q = \frac{1}{2Q(x)}.$$

Exercise. Show that $D_x L = \frac{1}{x}$ and $D_x E = E(x)$.

The Simple Graph I

The simple graph I is the simple graph to which the point P belongs only if the ordinate of P is the abscissa of P, so

$$I(x) = x \quad \text{for every number } x.$$

This is the straight line of slope 1 containing the point $(0, 0)$, and

$$D_x I = 1 \quad \text{for every number } x.$$

4

Combinations of Simple Graphs

Addition

The statement that $f + g$ is the *sum* of the simple graph f and the simple graph g means that there is a number common to the X-projection of f and the X-projection of g and $f + g$ is the simple graph whose X-projection is the common part of the X-projection of f and the X-projection of g such that, if x is in this common part, the ordinate of that point of $f + g$ whose abscissa is x is $f(x) + g(x)$, or

$$(f + g)(x) = f(x) + g(x).$$

Example. If f is the subset of the horizontal line $\underline{1}$ whose X-projection is the set of all nonpositive numbers and g the subset of the horizontal line $\underline{2}$ whose X-projection is the set of all nonnegative numbers, then $f + g$ is the simple graph containing only the point $(0, 3)$. In this case, f has slope 0 at $(0, 1)$, g has slope 0 at $(0, 2)$ but $f + g$ does not have slope at $(0, 3)$.

Problem. Suppose the simple graph f has slope m_1 at $(x, f(x))$, the simple graph g has slope m_2 at $(x, g(x))$, and each two vertical lines with $(x, f(x) + g(x))$ between them have between them a point of $f + g$ distinct from $(x, f(x) + g(x))$. Find the slope $D_x(f + g)$.

Multiplication

The statement that fg or $f \cdot g$ is the *product* of the simple graph f and the simple graph g means that there is a number common to the X-projection of f and the X-projection of g and that $f \cdot g$ is the simple graph defined for every number x common to the X-projection of f and the X-projection of g by

$$(f \cdot g)(x) = f(x) \cdot g(x).$$

Problem. Suppose f and g are as in the preceding problem. Find the slope $D_x(f \cdot g)$.

Bracket Multiplication

The statement that $f[g]$ (read f of g) is the *bracket product* f of g means there is a number x such that $g(x)$ is the abscissa of a point of f and $f[g]$ is defined for all such x by

$$f[g](x) = f(g(x)).$$

That is, the ordinate of that point of $f[g]$ whose abscissa is x is the ordinate of that point of f whose abscissa is $g(x)$.

There is a simple construction by which points of $f[g]$ may be located, shown in Figure 4.1.

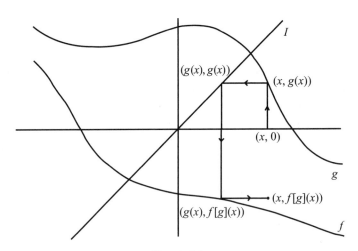

Figure 4.1

Starting with a point $(x, 0)$, where x is the abscissa of a point of g, draw a vertical line interval to the point $(x, g(x))$ of g, then a horizontal line interval to the point $(g(x), g(x))$ of the simple graph I, then a vertical line interval to the point $(g(x), f[g](x))$ of f and, finally, draw a horizontal line interval to the point $(x, f[g](x))$ on the vertical line $x \mid$.

Problem. Suppose the simple graph g has slope m_2 at $(x, g(x))$ and the simple graph f has slope m at $(g(x), f[g](x))$. Suppose each two vertical lines with $(x, f[g](x))$ between them have between them a point of $f[g]$ distinct from $(x, f[g](x))$. Find the slope

$$D_x f[g].$$

Division

The reciprocal $1/g$ of the simple graph g is $F[g]$, where F is the simple graph defined for every number x distinct from 0 by

$$F(x) = \frac{1}{x}.$$

The quotient f/g of the simple graph f by the simple graph g is $f \cdot (1/g)$. A number x is not in the X-projection of $1/g$ in case $g(x) = 0$.

Problem. Suppose the simple graph g has slope m at $(x, g(x))$ and $g(x) \neq 0$. Find the slope

$$D_x \frac{1}{g}.$$

Then, under suitable conditions on the simple graph f, find

$$D_x \frac{f}{g}.$$

Statements analogous to some of the axioms of the number system are true for the system of simple graphs, complicated somewhat by the fact that two simple graphs may not have the same X-projection. If X_f denotes the X-projection of the simple graph f, some of these statements are as follows:

(i) If each of f, g, and h is a simple graph and X_f, X_g, and X_h have a common part, then $(f + g) + h = f + (g + h)$.

(ii) $\underline{0}$ is a simple graph such that if f is a simple graph, $\underline{0} + f = f$.

(iii) If f is a simple graph, the simple graph $-f$ defined for every x in X_f by $(-f)(x) = -f(x)$ has the property that $f + (-f)$, denoted by $f - f$, is $\underline{0}f$.

(iv) If each of f and g is a simple graph and X_f and X_g have a common part, then $f + g = g + f$.

If k is a number, we ordinarily write kf for $\underline{k}f$ and $k + f$ for $\underline{k} + f$, for any simple graph f. If each of f and g is a simple graph and X_f and X_g have a common part, then $f + (-g)$ is written $f - g$ and called the *difference* f minus g.

The bracket product has interesting properties of which we mention a few. If f is a simple graph then

(i) $f[I] = I[f] = f$,

(ii) $f[\underline{1}] = \underline{f(1)}$, if 1 is in X_f,

(iii) $\underline{1}[f]$ is the subset of $\underline{1}$ whose X-projection is X_f,

(iv) $L[\underline{1}]$ is $\underline{0}$, and

(v) $\underline{1}[L]$ is the subset of $\underline{1}$ whose X-projection is the set of positive numbers.

If f is a simple graph, f^0 denotes $\underline{1}$ and, if n is a positive integer, $f^n = f \cdot f^{n-1}$.

The Derivative

The statement that f' is the *derivative* of the simple graph f means that f has slope at one of its points and that f' is the simple graph to which the point (x, y) belongs only if f has slope at $(x, f(x))$ and y is the slope of f at $(x, f(x))$.

Examples.

$$H' = -H^2,$$
$$(I^2)' = 2I,$$
$$L' = H,$$
$$E' = E,$$
$$Q' = \frac{1}{2Q}.$$

If k is a number, $\underline{k}' = \underline{0}$.

Under suitable conditions, e.g., if f, g, f' and g' have X-projection an interval $[a, b]$, we have

(i) $(f + g)' = f' + g'$,

(ii) $(fg)' = fg' + f'g$,

(iii) $(kf)' = kf'$, if k is a number.

Also, under suitable conditions

(iv) $(f[g])' = f'[g] \cdot g'$,

(v) $\left(\dfrac{1}{g}\right)' = \dfrac{-g'}{g^2}$,

and

(vi) $\left(\dfrac{f}{g}\right)' = \dfrac{gf' - fg'}{g^2}$.

Exercise. Show that if the simple graph f has slope at one of its points and n is a positive integer,

$$(f^n)' = nf^{n-1} \cdot f'.$$

If f is a simple graph, the absolute value of f, denoted by $|f|$, is the simple graph $Q[f^2]$. Its derivative is

$$|f|' = \frac{ff'}{|f|}.$$

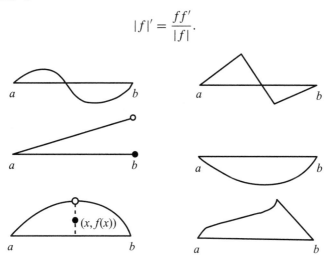

Figure 4.2

Problem. Each of the sketches in Figure 4.2 represents a simple graph f with X-projection the interval $[a, b]$, with $f(a)$ and $f(b)$ the number 0. Study the sketches and try to complete this statement of a theorem: If the simple graph f has X-projection the interval $[a, b]$, $f(a) = f(b) = 0$, and f has slope at each of its points, then. . . .

5

Theorems about Simple Graphs

We now state a theorem about simple graphs followed by a number of other theorems which, since they may be proved using the main theorem, are labeled *corollaries*. We suggest that the reader take the main theorem for granted temporarily and use it as an axiom to prove the corollaries. Then return to the main theorem from time to time and try to prove it.

Theorem. *If the simple graph f has X-projection the interval $[a, b]$ and has slope at each of its points, then there is a point P of f such that no point of f is higher than P and a point Q of f such that no point of f is lower than Q.*

Corollary 1. *If the simple graph f has X-projection the interval $[a, b]$, $f(a) = f(b) = 0$, and f has slope at each of its points, then there exists a number c between a and b such that f has slope 0 at $(c, f(c))$, or*

$$D_c f = 0.$$

Corollary 2. *Suppose each of f and g is a simple graph with X-projection the interval $[a, b]$ each having slope at each of its points with $f(a) = g(a)$ and $f(b) = g(b)$. Then there exists a number c between a and b such that*

$$D_c f = D_c g.$$

If g is a straight line, this becomes

$$\frac{f(b) - f(a)}{b - a} = f'(c)$$

or

$$f(b) - f(a) = (b - a) \cdot f'(c).$$

Corollary 3. *If the simple graph f has X-projection the interval $[a, b]$ and the slope of f is 0 at each of its points, then $f(x) = f(a)$ for every number x in $[a, b]$.*

Corollary 4. *Suppose f is a simple graphs whose X-projection is the interval $[a, b]$ having a slope at each of its points and there is a number c between a and b such that*

$$f'(x) \begin{cases} < 0, & \text{if } a \leq x < c, \\ = 0, & \text{if } x = c, \\ > 0, & \text{if } c < x \leq b. \end{cases}$$

Then the point $(c, f(c))$ is lower than any other point of f.

Note. There is an analogous condition for a point to be the highest point of f.

Corollary 5. *Suppose each of f and g is a simple graph with X-projection the interval $[a, b]$ having slope at each of its points so that $f(a) = g(a)$, $f(b) > g(b)$ and $D_b f < D_b g$. Then there exists a number c between a and b such that $D_c f = D_c g$.*

Corollary 6. *Suppose each of f and g is a simple graph with X-projection $[a, b]$ having slope at each of its points so that $f(a) = g(a)$ and, if x is a number between a and b, $D_x f > D_x g$. Then $f(b) > g(b)$.*

Corollary 7. *Suppose the simple graph f has X-projection the interval $[a, b]$ and has slope at each of its points. If m is a number between $D_a f$ and $D_b f$, then there exists a number c between a and b such that $D_c f = m$.*

The Conjectured Computation Formula

In an earlier chapter, we conjectured that if c is a number greater than 1 and n a positive integer, then

$$L(c) = \frac{c-1}{c} + \frac{1}{2}\left(\frac{c-1}{c}\right)^2 + \cdots + \frac{1}{n}\left(\frac{c-1}{c}\right)^n + k_n,$$

$$0 < k_n < \frac{1}{n+1}\left(\frac{c-1}{c}\right)^{n+1} \cdot c.$$

To establish this, we suggest that the reader adopt a more general point of view and consider the simple graph K_n defined by

$$K_n = L - (1 - H) - \frac{1}{2}(1 - H)^2 - \cdots - \frac{1}{n}(1 - H)^n,$$

so that the number k_n is $K_n(c)$. Two applications of Corollary 6 with suitable choices of f and g in each case will prove the conjecture.

The Addition Formulas for L and E

We had the formulas

$$L(cx) = L(c) + L(x)$$

and

$$E(c + x) = E(c)E(x),$$

the first being true for $c > 0$ and $x > 0$ and the second for any number c and any number x. These may be called *addition formulas*. They may be proved by adopting a more general point of view and using Corollary 3. The first formula may be stated as follows: the simple graph f defined by

$$f = L[cI] - L - L(c)$$

is a subset of $\underline{0}$. The second may be stated as follows: The simple graph f defined by

$$f = \frac{E[c + I]}{E(c) \cdot E}$$

is the horizontal line $\underline{1}$.

The Simple Graphs S and C

These are defined as follows:

$$S = \frac{1}{2}\left\{E - \frac{1}{E}\right\} \quad \text{and} \quad C = \frac{1}{2}\left\{E + \frac{1}{E}\right\}.$$

Since $\frac{1}{E(x)} = E(-x)$ for every number x, we see that the point (x, y) belongs to the simple graph $1/E$ only if the point $(-x, y)$ belongs to E. To sketch S and C, sketch E and $1/E$ and then take half the difference and then half their sum.

Exercise.

(i) Show that $S' = C$ and $C' = S$.

(ii) Show that $C^2 - S^2 = \underline{1}$.

(iii) Show that if each of x and c is a number, then

$$S(x + c) = S(x)\, C(c) + C(x)\, S(c)$$

and

$$C(x + c) = C(x)\, C(c) + S(x)\, S(c).$$

The derivative of f', $(f')'$, is denoted by f'' and called the *second* derivative of f.

(iv) Show that $S'' = S$ and $C'' = C$.

(v) Sketch the simple graphs K and T defined by

$$K = \frac{1}{C} \quad \text{and} \quad T = \frac{S}{C}.$$

(vi) Show that $T' = K^2$ and $K' = -KT$. Also, $T' = 1 - T^2$.

(vii) Show that if each of x and y is a number, then

$$T(x + y) = \frac{T(x) + T(y)}{1 + T(x)\, T(y)}.$$

(viii) Show that if x and y are numbers and $x < y$, then $T(x) < T(y)$.

(ix) Show that every point of the simple graph T is between the horizontal lines $-\underline{1}$ and $\underline{1}$. Moreover, show that 1 is the least positive number k such that each point of T is between $-\underline{k}$ and $\underline{k}$.

Exercise. A denotes the simple graph to which the point (x, y) belongs only if the point (y, x) belongs to T. The X-projection of A is the set of all numbers between -1 and 1.

(i) Show that

$$D_x \, \mathcal{A} = \frac{1}{1 - x^2}, \ -1 < x < 1.$$

(ii) Show that if c is a number between -1 and 1, the simple graph

$$\mathcal{A} + \mathcal{A}(c) - \mathcal{A}\left[\frac{I + c}{1 + Ic}\right]$$

is a subset of $\underline{0}$, so that, if x is a number between -1 and 1,

$$\mathcal{A}(x) + \mathcal{A}(c) = \mathcal{A}\left(\frac{x + c}{1 + xc}\right).$$

(iii) Show that

$$\mathcal{A} = \frac{1}{2} L\left[\frac{1 + I}{1 - I}\right].$$

A Point Set which is not a Simple Graph

Suppose G denotes the point set to which (x, y) belongs only if $y^2 = x^2(1 - x^2)$. There are two points of G having the same abscissa, so that G is not a simple graph. The points of G constitute two simple graphs,

$$IQ[1 - 1^2] \quad \text{and} \quad -IQ[1 - I^2].$$

Denote the first of these by f.
    The X-projection of f is $[-1, 1]$.

Exercise. Calculate f', find the highest point of f, the lowest point of f, and sketch f and then G. Show that

$$f'' = \frac{I(2I^2 - 3)}{(1 - I^2)Q[1 - I^2]}.$$

Note that $f''(x)$ is positive, 0, or negative according as $-1 < x < 0$, $x = 0$, or $0 < x < 1$. What does this say about the shape of f? Does f have slope 1.001 at any of its points?

Inequalities

Theorem. *If the simple graph f has X-projection the interval $[a, b]$, if f'', the second derivative of f, has X-projection $[a, b]$, and if $f''(x) > 0$ for every number x between a and b, then every point of f with abscissa between a and* *b is below the line containing the points $(a, f(a))$ and $(b, f(b))$. That is*

$$f(x) < \frac{f(b) - f(a)}{b - a}(x - a) + f(a), \qquad a < x < b.$$

Suppose k is a positive number distinct from 1 and f is the simple graph defined for every number x by

$$f(x) = k^x.$$

Thus, $f = E[IL(k)]$. Then $f' = E'[IL(k)] \cdot \{IL(k)\}' = f \cdot L(k)$ and $f'' = f \cdot L^2(k)$, so that $f''(x) > 0$ for every number x. Thus, if we take $[a, b]$ to be $[0, 1]$ we have by the Theorem

$$k^x < (k - 1) \cdot x + 1, \qquad 0 < x < 1,$$

or

$$k^x < k \cdot x + (1 - x), \qquad 0 < x < 1.$$

If a and b are positive numbers and $k = a/b$, the inequality may be written

$$a^x \cdot b^{1-x} < a \cdot x + b \cdot (1 - x), \qquad 0 < x < 1.$$

If $x = (1/2)$ this becomes

$$\sqrt{a \cdot b} < \frac{a + b}{2}.$$

That is, the geometric mean of two positive numbers is less than their arithmetic mean.

If each of a, b, and c is a positive number with $b \neq c$ and each of x, y, and z is a positive number and $x + y + z = 1$, then

$$a^x \cdot b^y \cdot c^z = a^x \cdot \left\{ b^{\frac{y}{y+z}} \cdot c^{\frac{z}{y+z}} \right\}^{y+z} \leq a \cdot x + \left\{ b^{\frac{y}{y+z}} \cdot c^{\frac{z}{y+z}} \right\} \cdot (y + z)$$

and therefore

$$a^x \cdot b^y \cdot c^z < x \cdot a + y \cdot b + z \cdot c.$$

If $x = y = z = \frac{1}{3}$, this gives

$$\sqrt[3]{abc} < \frac{a + b + c}{3}.$$

The reader may follow this up and find many other inequalities.

6

The Simple Graphs of Trigonometry

Following the pattern by which the simple graphs L and E were developed from H, simple graphs A and T may be developed from Ω, where (see Figure 6.1)

$$\Omega = \frac{1}{1 + I^2}.$$

There are enough differences and similarities between the two developments to make the work interesting. We shall give a broad outline, leaving the details to the reader.

If $[a, b]$ is an interval, we denote by $[\Omega; a, b]$ the point set to which (x, y) belongs only if x belongs to $[a, b]$ and y is 0, y is $\frac{1}{1+x^2}$, or y is a number between 0 and $\frac{1}{1+x^2}$. An inner sum for $[\Omega; a, b]$ is the sum of the areas of finitely many nonoverlapping rectangular intervals included in $[\Omega; a, b]$, their bases filling up $[a, b]$ and each one as high as possible. Figure 6.1 shows three such rectangular intervals.

The area of $[\Omega; a, b]$, denoted by $\int_a^b \Omega$, is the least number that no inner sum for $[\Omega; a, b]$ exceeds. A denotes the simple graph whose X-projection is the set of all numbers defined by

$$A(x) = \begin{cases} -\int_x^0 \Omega, & \text{if } x < 0, \\ 0, & \text{if } x = 0, \\ \int_0^x \Omega, & \text{if } x > 0. \end{cases}$$

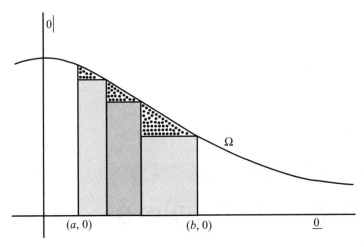

Figure 6.1

If (x, y) is a point of A, then $(-x, -y)$ is a point of A.

If $[p, q]$ is an interval of nonnegative numbers, we have the approximation

$$\int_p^q \Omega \doteq \frac{q - p}{2} \cdot \frac{p^2 + q^2 + 2}{(1 + p^2)(1 + q^2)}$$

with error less than

$$\frac{(q - p)^2}{2} \cdot \frac{q + p}{(1 + p^2)(1 + q^2)}.$$

As with L, this formula may be used to compute a table of values

x	$A(x)$
0	0
.1	.0995
.2	.1971
.3	.2911
.4	
.5	
.6	
.7	
.8	
.9	
1.0	

by means of the following theorem.

Theorem. *If* $[a, b]$ *is an interval and c a number between a and b, then*

$$\int_a^b \Omega = \int_a^c \Omega + \int_c^b \Omega$$

and

$$A(b) - A(a) = \int_a^b \Omega.$$

It follows that, if x and y are nonnegative numbers or if x and y are nonpositive numbers, then $A(y) - A(x)$ is between $(y - x) \Omega(x)$ and $(y - x) \Omega(y)$. That is, there exists a number t between 0 and 1 such that

$$A(y) - A(x) = (y - x)\{(1 - t) \cdot \Omega(x) + t \cdot \Omega(y)\}$$

or

$$A(y) - A(x) = (y - x) \cdot \Omega(x) + (y - x) \cdot t \cdot \{\Omega(y) - \Omega(x)\}.$$

Consequently, $D_x A = \Omega(x)$ or $A' = \Omega$.

Problem. Show that if x is a number then $-3 < A(x) < 3$.

We denote by $\pi/2$ the least positive number k such that the simple graph A is between the horizontal lines $-\underline{k}$ and $\underline{k}$. See Figure 6.2.

If x and y are numbers with $x < y$, then $A(x) < A(y)$ and the Y-projection of A is the set of all numbers between $-\pi/2$ and $\pi/2$.

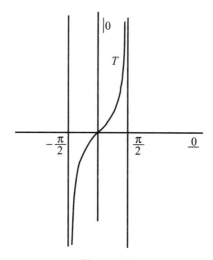

Figure 6.2

The point set to which (x, y) belongs only if (y, x) belongs to A is a simple graph which we denote by T. The X-projection of T is the set of all numbers between $-\pi/2$ and $\pi/2$. The derivative of T is $1 + T^2$:

$$T' = 1 + T^2.$$

Addition Formulas for A

There are some analogies between the graphs $\mathcal{A}$ and A. For instance, we have

$$D_x\mathcal{A} \;=\; \frac{1}{1 - x^2} \qquad \text{and} \qquad D_x A = \frac{1}{1 + x^2}$$

and

$$T' \;=\; 1 - T^2 \qquad \text{and} \qquad T' = 1 + T^2.$$

If each of x and c is a number between -1 and 1, then

$$\mathcal{A}(x) + \mathcal{A}(c) = \mathcal{A}\left(\frac{x + c}{1 + xc}\right).$$

This is analogous to the formula

$$L(x) + L(c) = L(xc) \quad (x > 0, c > 0).$$

A study of the simple graph A shows that there are numbers x and c such that $A(x) + A(c)$ is not the ordinate of any point of A, as when $x > 1$ and $c > 1$. On the other hand, if $xc < 1$ it would seem that $A(x) + A(c)$ is the ordinate of some point of A. We conjecture, by analogy with the addition formula for $\mathcal{A}$, that if each of x and c is a number and $xc < 1$, then

$$A(x) + A(c) = A\left(\frac{x + c}{1 - xc}\right).$$

Phrased differently, if c is a number, the simple graph f defined by

$$f = A + A(c) - A\left[\frac{I + c}{1 - Ic}\right],$$

whose X-projection is the set of all numbers x for which $cx \neq 1$, has slope 0 at each of its points. Since $f(0) = 0$, it follows that $f(x) = 0$ if $cx < 1$, which is our conjecture; and there exists a number k such that $f(x) = k$ if $cx > 1$.

Exercise. Show that if x is a number, then

(i) $A(x) + A\left(\dfrac{1}{x}\right) = \begin{cases} \dfrac{\pi}{2}, & \text{if } x > 0, \\ -\dfrac{\pi}{2}, & \text{if } x < 0, \end{cases}$

(ii) $A(x) + A(1) = A\left(\dfrac{1+x}{1-x}\right) \quad \text{if } x < 1,$

(iii) $A(x) - A(1) = A\left(\dfrac{x-1}{x+1}\right) \quad \text{if } x > -1,$

(iv) $A(x) + A(c) - A\left(\dfrac{x+c}{1-xc}\right) = \begin{cases} \pi, & \text{if } xc > 1 \text{ and } x > 0, \\ -\pi & \text{if } xc > 1 \text{ and } x < 0. \end{cases}$

We include the formula previously stated:

(v) $A(x) + A(c) = A\left(\dfrac{x+c}{1-xc}\right), \quad \text{if } xc < 1.$

Show that, if each of x and y is a number between $-\pi/2$ and $\pi/2$ such that $x + y$ is between $-\pi/2$ and $\pi/2$, then

$$T(x + y) = \frac{T(x) + T(y)}{1 - T(x)T(y)}.$$

Problem. Show that $T(\pi/4) = 1$.

Computation Problem for A

If x is a number, $A(-x) = -A(x)$ and $A(0) = 0$, so to compute $A(x)$ we need consider only positive values of x.

If c is a positive number,

$$A(c) = \int_0^c \Omega > c \cdot \frac{1}{1+c^2},$$

so that

$$A(c) = \frac{c}{1+c^2} + \text{(a positive number)}.$$

Let A_1 denote the simple graph defined by

$$A_1 = A - \frac{I}{1+I^2}.$$

Then this positive number is $A_1(c)$. We find that

$$A_1' = \frac{2I^2}{(1+I^2)^2},$$

so that if $0 < x < c$, $D_x A_1 > D_x \underline{0}$. Since $A_1(0) = \underline{0}(0)$, it follows from an earlier theorem that $A_1(c) > 0$. We next seek a simple graph g such that $A_1(0) = g(0)$ and $D_x A_1 < D_x g$, for $0 < x < c$. After some experimentation, we find that such a simple graph g is

$$C \cdot \frac{I^2}{1 + I^2},$$

so

$$0 < A_1(c) < \frac{c}{1 + c^2} \cdot c^2$$

$$A(c) = \frac{c}{1 + c^2} + A_1(c).$$

The fact that $A(x) = -A(-x)$ for every number x, so that $A_1(x) = -A_1(-x)$ for every number x, might lead us to try next something such as

$$A(c) = \frac{c}{1 + c^2} + \frac{ac^3}{(1 + c^2)^2} + A_2(c),$$

where a is a positive number. (If $a < 0$, the new formula wouldn't be as good as the one already found.) If

$$A_2 = A - \frac{I}{1 + I^2} + \frac{aI^3}{(1 + I^2)^2} = A_1 - \frac{aI^3}{(1 + I^2)^2},$$

then

$$A_2' = A_1' - \frac{(1 + I^2)^2 \cdot 3aI^2 - aI^3 \cdot 2(1 + I^2) \cdot 2I}{(1 + I^2)^4} = \frac{(2 - 3a)I^2 + (2 + a)I^4}{(1 + I^2)^3}.$$

This suggests that we choose a to be $2/3$, so that

$$A_2' = \frac{8}{3} \cdot \frac{I^4}{(1 + I^2)^3}.$$

Then $A_2(c) > 0$ and after some experimentation we find that a simple graph g for which $g(0) = A_2(0)$ and $D_x A_2 < D_x g$ for $0 < x < c$ is

$$\frac{2}{3} c \cdot \left(\frac{I^2}{(1 + I^2)} \right)^2.$$

Consequently,

$$0 < A_2(c) < \frac{2}{3} \cdot \frac{c^3}{(1 + c^2)^2} \cdot c^2,$$

$$A(c) = \frac{c}{1 + c^2} + \frac{2}{3} \cdot \frac{c^3}{(1 + c^2)^2} + A_2(c).$$

We have now found

$$A_1 = A - \frac{I}{1 + I^2} \quad \text{and} \quad A_1' = \frac{2}{1} \cdot \frac{I^2}{(1 + I^2)^2}$$

$$A_2 = A_1 - \frac{2}{3} \cdot \frac{I^3}{(1 + I^2)^2} \quad \text{and} \quad A_2' = \frac{2}{1} \cdot \frac{4}{3} \cdot \frac{I^4}{(1 + I^2)^3}$$

and we conjecture that

$$A_3 = A_2 - \frac{2}{3} \cdot \frac{4}{5} \cdot \frac{I^5}{(1 + I^2)^3} \quad \text{and} \quad A_3' = \frac{2}{1} \cdot \frac{4}{3} \cdot \frac{6}{5} \cdot \frac{I^8}{(1 + I^2)^4},$$

and so on, and if n is a positive integer, that

$$A(c) = \frac{c}{1 + c^2} + \frac{2}{3} \cdot \frac{c^3}{(1 + c^2)^2} + \cdots + \frac{2}{3} \cdot \frac{4}{5} \cdot \ldots \cdot \frac{2n}{2n + 1} \cdot \frac{c^{2n+1}}{(1 + c^2)^{n+1}}$$
$$+ A_n(c),$$

where

$$0 < A_n(c) < \frac{2}{3} \cdot \frac{4}{5} \cdot \ldots \cdot \frac{2n}{2n + 1} \cdot \frac{c^{2n+1}}{(1 + c^2)^{n+1}} \cdot c^2.$$

This is a true statement.

Taking c to be 1, we find that $A(1) = .7854$ to four decimal places, and consequently π, or $4A(1)$, is 3.1416 to four decimal places.

The Simple Graphs K, C, S, and B

The simple graph K whose X-projection is the set of all numbers between $-\pi/2$ and $\pi/2$ is defined by

$$K = Q[1 + T^2].$$

The simple graph C whose X-projection is the interval $[-\pi/2, \pi/2]$ is defined by

$$C(x) = \begin{cases} 0, & x = -\dfrac{\pi}{2}, \\ \dfrac{1}{K(x)}, & -\dfrac{\pi}{2} < x < \dfrac{\pi}{2}, \\ 0, & x = \dfrac{\pi}{2}. \end{cases}$$

The simple graphs S whose X-projection is the interval $[-\pi/2, \pi/2]$ is defined by

$$S(x) = \begin{cases} -1, & x = -\dfrac{\pi}{2}, \\[2mm] C(x)T(x), & -\dfrac{\pi}{2} < x < \dfrac{\pi}{2}, \\[2mm] 1, & x = \dfrac{\pi}{2}. \end{cases}$$

Exercise. Show that

(i) $T' = K^2$,

(ii) $K' = KT$,

(iii) $S' = C$ and $C' = -S$,

(iv) $C^2 + S^2$ is a subset of $\underline{1}$,

(v) $T = \dfrac{S}{C}$,

and

(vi) if each of u and v is a number in the interval $[-\pi/2, \pi/2]$ such that $u + v$ is in this interval, then $S(u + v) = S(u)C(v) + C(u)S(v)$ and $C(u + v) = C(u)C(v) - S(u)S(v)$.

The Y-projection of S is the interval $[-1, 1]$ and if x and y are numbers in this interval such that $x < y$, then $S(x) < S(y)$. It follows that the point set B to which (x, y) belongs only if (y, x) belongs to S is a simple graph whose X-projection is $[-1, 1]$. The Y-projection of B is, of course, $[-\pi/2, \pi/2]$.

If x is a number between -1 and 1,

$$D_x B = \frac{1}{Q[1 - x^2]}.$$

That is,

$$B' = \frac{1}{Q[1 - I^2]}.$$

Extension of S and C to the Set of all Numbers

We now extend the simple graphs S and C so that their X-projections shall be the set of all numbers by requiring that the formulas

$$S(u + v) = S(u)C(v) + C(u)S(v),$$
$$C(u + v) = C(u)C(v) - S(u)S(v),$$

shall be true for all numbers. To do this it is sufficient to require that

$$S(u + 2\pi) = S(u)$$
$$C(u + 2\pi) = C(u)$$

for every number u. We then extend the definition of K by

$$K = \frac{1}{C}$$

and the definition of T by

$$T = \frac{S}{C}.$$

The formulas

$$S' = C, \quad C' = -S, \quad T' = K^2, \quad \text{and} \quad K' = KT$$

are true for the extended simple graphs. Also,

$$C^2 + S^2 = 1 \quad \text{and} \quad 1 + T^2 = K^2.$$

We leave to the reader the project of constructing detailed proofs of the statements made in the preceding outline. One of the main problems is to show that

$$D_x C = -S(x) \quad \text{and} \quad D_x S = C(x)$$

are true if $x = -(\pi/2)$ and $x = \pi/2$.

Exercise.

   (i) Find the derivatives of $L[|K|]$ and $L[|K + T|]$.

   (ii) Show that

$$S[A] = \frac{I}{Q[1 + I^2]} \quad \text{and} \quad C[A] = \frac{1}{Q[1 + I^2]}.$$

   (iii) Show that $C[3I] = 4C^3 - 3C$ and use this to find $S(x), C(x), T(x)$, and $K(x)$ if $x = \pi/6$ and $x = \pi/3$.

   (iv) Find $S(x), C(x)$, and $K(x)$ if $x = \pi/4$.

Connection with Triangles

Theorem. *If r is a positive number and each of x and y is a number such that* $x^2 + y^2 = r^2$, *then there exists only one number* θ *such that* $0 \le \theta < 2\pi$ *and*

$$S(\theta) = \left(\frac{y}{r}\right) \quad and \quad C(\theta)\left(\frac{x}{r}\right).$$

Definition. The number θ of the preceding theorem is called the argument of the point (x, y). That is, if (x, y) is a point such that $x^2 + y^2$ is the positive number r^2 $(r > 0)$, then the number θ such that $0 \le \theta < 2\pi$, $S(\theta) = (y/r)$, and $C(\theta) = (x/r)$ is the argument of (x, y).

If (x, y) is a point distinct from $(0, 0)$ and t is a positive number, then the argument of (x, y) is the argument of (tx, ty). Since $T(\theta) = S(\theta)/C(\theta)$, then $T(\theta) = y/x$, provided $x \ne 0$. Thus, $T(\theta)$ is the slope of the straight line containing $(0, 0)$ and (x, y). The number r is the distance from $(0,0)$ to (x, y). Thus, if $x > 0$ and $y > 0$, $S(\theta)$, $C(\theta)$, $T(\theta)$, and $K(\theta)$ are ratios of lengths of sides of a triangle:

$$S(\theta) = \frac{y}{r}, \quad C(\theta) = \frac{x}{y}, \quad T(\theta) = \frac{y}{x}, \quad K(\theta) = \frac{r}{x}.$$

See Figure 6.3. There are two more ratios,

$$\frac{1}{S(\theta)} \quad and \quad \frac{1}{T(\theta)}$$

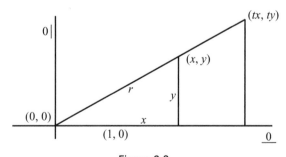

Figure 6.3

Property S

The sketches in Figure 6.4 show simple graphs with one point designated by the letter P. Those simple graphs marked (Y) have property S at the point P

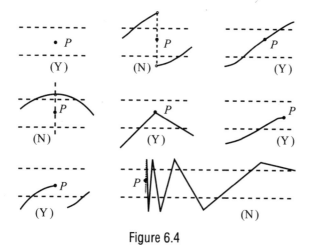

Figure 6.4

and those marked (N) do not have property S at P. See if you can describe this property in terms of pairs of horizontal and vertical lines.

7

The Integral

The statement that the simple graph f *is bounded on the interval* $[a, b]$ means that the X-projection of f includes $[a, b]$ and that there exist horizontal lines α and β such that every point of f whose abscissa is in $[a, b]$ is between α and β.

Suppose that the simple graph f is bounded on $[a, b]$. The statement that $_iS$ *is an inner sum for* f *on* $[a, b]$ means that there exists a finite collection D of nonoverlapping intervals filling up $[a, b]$ such that, if the length of each interval in D is multiplied by the largest number which exceeds the ordinate of no point of f whose abscissa is in that interval, then the sum of all the products so formed is $_iS$. The inner sum may be described as *based* on D and designated as $_iS_D$. The statement that $_oS$ *is an outer sum for* f *on* $[a, b]$ means that there exists a finite collection D of nonoverlapping intervals filling up $[a, b]$ such that, if the length of each interval in D is multiplied by the smallest number which the ordinate of no point of f with abscissa in that interval exceeds, then the sum of all the products so formed is $_oS$ or $_oS_D$ (See Figure 7.1).

Since f is bounded on $[a, b]$, there is a number that no inner sum for f on $[a, b]$ exceeds. The *least* number that no inner sum for f on $[a, b]$ exceeds is denoted by

$$_i\!\int_a^b f$$

and called the *inner integral from a to b of* f. Similarly, there is a number that exceeds no outer sum for f on $[a, b]$. The *largest* number that exceeds no outer

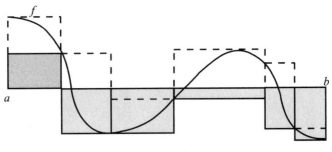

Figure 7.1

sum for f on $[a, b]$ is denoted by

$$_o\int_a^b f$$

and called the *outer integral from a to b of* f.

Theorem. *Suppose the simple graph* f *is bounded on* $[a, b]$ *and each of* α *and* β *is a number such that if* x *is a number in* $[a, b]$, *then* $\alpha \le f(x) \le \beta$. *If* $_iS$ *is an inner sum for* f *on* $[a, b]$ *and* $_oS$ *an outer sum for* f *on* $[a, b]$, *then*

$$(b - a) \cdot \alpha \le {}_iS \le {}_i\int_a^b f \le {}_o\int_a^b {}_oS \le (b - a) \cdot \beta.$$

There exists a simple graph f that is bounded on the interval $[0, 1]$ such that $_i\int_0^1 f = 0$ and $_o\int_0^1 f = 1$.

If $[a, b]$ is an interval of positive numbers, $_i\int_a^b H = {}_o\int_a^b H$, and if $[a, b]$ is an interval, $_i\int_a^b \Omega = {}_o\int_a^b \Omega$.

Definition. The statement that the simple graph f is integrable on the interval $[a, b]$ means that f is bounded on $[a, b]$ and $_i\int_a^b f = {}_o\int_a^b f$. If the simple graph f is integrable on $[a, b]$, the number $_i\int_a^b -f$, which is $_o\int_a^b f$, is denoted by $\int_a^b f$ and called *the integral from a to b of* f.

Theorem. *Suppose the simple graph* f *is bounded on the interval* $[a, b]$ *and* c *is a number between* a *and* b. *Then*

$$_i\int_a^b f = {}_i\int_a^c f + {}_i\int_c^b f$$

and

$$\int_{o}^{b} f = \int_{o}^{c} f + \int_{o}^{b} f.$$

If

$$g(x) = \begin{cases} 0, & x = a, \\ {}_i\int_a^x f, & a < x \le b, \end{cases}$$

and $$h(x) = \begin{cases} 0, & x = a, \\ {}_o\int_a^x f, & a < x \le b, \end{cases}$$

then if $[x, y]$ *is a subinterval of* $[a, b]$*, we have*

$$g(y) - g(x) = {}_i\int_x^y f \quad \text{and} \quad h(y) - h(x) = {}_o\int_x^y f.$$

Question. Is it true that $g' = f$ and $h' = f$? (We had $L' = H$ and $A' = \Omega$.)

Property S

The statement that the simple graph f has property S at the point P means that P is a point of f and if $\underline{\alpha}$ and $\underline{\beta}$ are horizontal lines with P between them, there exist vertical lines $h|$ and $k|$ with P between them such that every point of f between $h|$ and $k|$ is between $\underline{\alpha}$ and $\underline{\beta}$.

Theorem. *Suppose* $f, g,$ *and* h *are as in the preceding theorem, and* x *is a number in* $[a, b]$ *such that* f *has property* S *at the point* $(x, f(x))$*. Then*

$$D_x g = f(x) \text{ and } D_x h = f(x).$$

If f has X-projection $[a, b]$ and has property S at each of its points, then $g' = f$ and $h' = f$ and $g = h$, so that

$$\int_{i}^{b} f = \int_{o}^{b} f.$$

I.e., f is integrable on $[a, b]$.

If the simple graph f has slope at the point P, then f has property S at P, but there exists a simple graph f having property S at the point P which does not have slope at P.

If the reader has succeeded in proving the theorem on page 37, it may be that the proof can be modified to establish the following more general theorem.

If not, we suggest working on the following theorem, which includes the theorem on page 37 as a corollary.

Theorem. *If the simple graph f has X-projection the interval $[a, b]$ and has property S at each of its points, then there exists a point P of f such that no point of f is higher than P and a point Q of f such that no point of f is lower than Q.*

Corollary 1. *If the simple graph f has X-projection the interval $[a, b]$ and has property S at each of its points, then f is integrable on $[a, b]$.*

Corollary 2. *If the simple graph f has X-projection the interval $[a, b]$ and has property S at each of its points, then there exists a simple graph g with X-projection $[a, b]$ such that $g' = f$.*

Corollary 3. *Suppose f, P, and Q are as in the theorem, and $\underline{h}$ is a horizontal line between P and Q. There exists a point of f belonging to $\underline{h}$.*

The corollaries to the main theorem on page 37 may be somewhat improved. We state here an improved version of the first corollaries and suggest that the reader try to improve the others.

Theorem. *If the simple graph f has X-projection the interval $[a, b]$ and has property S at each of its points, if $f(a) = f(b) = 0$, and if f has slope at each of its points with abscissa between a and b, then there exists a number c between a and b such that*

$$D_c f = 0.$$

Problem. Under suitable hypotheses upon f and g, investigate the question of property S for $f + g$, $f \cdot g$, $f[g]$, and f/g.

Properties of Integrals

(i) If f is a simple graph such that f' is bounded on the interval $[a, b]$, then

$$\int_{i}{}_a^b f' \le f(b) - f(a) \le \int_{o}{}_a^b f',$$

so that, if f' is integrable on $[a, b]$,

$$\int_a^b f' = f(b) - f(a).$$

(ii) If each of f and g is a simple graph that is integrable on $[a, b]$ and if k is a number, then $f + g$ and kf are integrable on $[a, b]$ and

$$\int_a^b (f + g) = \int_a^b f + \int_a^b g \quad \text{and} \quad \int_a^b kf = k \cdot \int_a^b f.$$

(iii) If each of f and g is integrable on $[a, b]$, then $f \cdot g$ is integrable on $[a, b]$.

(iv) If each of f, g, f', and g' is integrable on $[a, b]$, then

$$\int_a^b fg' = \int_a^b (fg)' - \int_a^b gf' = f(b)g(b) - f(a)g(a) - \int_a^b gf'.$$

(v) If f is integrable on $[a, b]$ and c is a number between a and b, then f is integrable on $[a, c]$ and on $[c, b]$ and

$$\int_a^b f = \int_a^c f + \int_c^b f.$$

Evaluation of Integrals

To develop facility in evaluation of integrals, we can use properties (i)–(v) to write an integral in different forms. By (i), any integral that can be written as $\int_a^b f'$ can be evaluated as $f(b) - f(a)$.

Examples.

$$\int_a^b C = \int_a^b S' = S(b) - S(a), \int_a^b S = \int_a^b (-C)' = C(a) - C(b).$$

If f is a simple graph whose X-projection includes the interval $[a, b]$, $[f; a, b]$ denotes the point set to which (x, y) belongs only if x is in $[a, b]$ and $y = 0$, $y = f(x)$ or y is a number between 0 and $f(x)$. If $f(x) \geq 0$ for every number x in $[a, b]$, the area of $[f; a, b]$ is *defined* to be the integral $\int_a^b f$. For instance, the area of $[S; 0, \pi/2]$ is $\int_0^{\pi/2} S = 2$.

Exercise 1. Evaluate

(i) $\displaystyle\int_a^b k$, i.e., $\displaystyle\int_a^b (kI)'$,   (iii) $\displaystyle\int_a^b I^2$,   (v) $\displaystyle\int_a^b E$,

(ii) $\displaystyle\int_a^b I$,   (iv) $\displaystyle\int_a^b I^3$,   (vi) $\displaystyle\int_a^b S$,

(vii) $\displaystyle\int_a^b C,$ (ix) $\displaystyle\int_a^b KT,$ (xi) $\displaystyle\int_a^b \frac{1}{I}.$

(viii) $\displaystyle\int_a^b K^2,$ (x) $\displaystyle\int_a^b \frac{1}{1+I^2},$

The interval $[a, b]$ must be subject to a certain restriction in (viii), (ix), and (xi).

The next examples illustrate the use of property (iv) to write integrals in different forms.

$$\int_a^b S^2 = \int_a^b SS = \int_a^b S \cdot (-C)' = \int_a^b (-SC)' - \int_a^b (-C)S'$$
$$= \int_a^b (-SC)' + \int_a^b C^2 = \int_a^b (-SC)' + \int_a^b (1 - S^2),$$

or

$$2 \int_a^b S^2 = \int_a^b (-SC)' + \int_a^b I' = \int_a^b (I - SC)',$$

or

$$\int_a^b S^2 = \int_a^b \left\{ \frac{I - SC}{2} \right\}'.$$

For instance, the area of $[S^2; 0, \frac{\pi}{2}]$ is $\int_0^{\pi/2} S^2$ or $\frac{\pi}{4}$.

Next,

$$\int_a^b S^3 = \int_a^b S^2 S = \int_a^b S^2 \cdot (-C)' = \int_a^b (-CS^2)' - \int_a^b (-C)(S^2)'$$
$$= \int_a^b (-CS^2)' + \int_a^b C \cdot 2SS' = \int_a^b (-CS^2)' + 2 \int_a^b C^2 S$$
$$= \int_a^b (-CS^2)' + 2 \int_a^b (1 - S^2) \cdot S,$$

or

$$3 \int_a^b S^3 = \int_a^b (-CS^2)' + 2 \int_a^b (-C)' = \int_a^b \{-CS^2 - 2C\}',$$

or

$$\int_a^b S^3 = \int_a^b \left\{ \frac{-CS^2 - 2C}{3} \right\}'.$$

Thus, $\int_0^{\pi/2} S^3 = \frac{2}{3}.$

Exercise 2. If n is a nonnegative integer, let k_n denote $\int_0^{\pi/2} S^n$. Then

$$k_0 = \frac{\pi}{2}, \qquad k_2 = \frac{\pi}{2} \cdot \frac{1}{2},$$

$$k_1 = 1, \qquad k_3 = 1 \cdot \frac{2}{3}.$$

Show that

$$k_n = \frac{n-1}{n} \cdot k_{n-2}, \quad n = 2, 3, 4, \ldots .$$

Exercise 3. Evaluate

(i) $\int_a^b L$, i.e., $\int_a^b L \cdot I'$, $(0 < a < b)$,

(ii) $\int_a^b \frac{I}{1+I^2}$, i.e., $\frac{1}{2} \int_a^b \{L[1 + I^2]\}'$,

(iii) $\int_a^b A$, i.e., $\int_a^b A \cdot I'$,

(iv) $\int_a^b T$,

(v) $\int_a^b K$.

The interval $[a, b]$ must be subject to restrictions in (i), (iv), and (v).

Exercise 4. Evaluate

(i) $\int_a^b K^3$,

(ii) $\int_a^b T^3$,

(iii) $\int_a^b L^2$,

(iv) $\int_a^b I S$.

The interval $[a, b]$ must be restricted in some of these.

Examples. Since we see by inspection that

$$\frac{1}{(I+1)(I+2)} = \frac{1}{I+1} - \frac{1}{I+2},$$

then for any interval not containing -1 or -2,

$$\int_a^b \frac{1}{(I+1)(I+2)} = \int_a^b \{L[|I+1|] - L[|I+2|]\}'.$$

Similarly,

$$\frac{1}{(I+1)(I+2)(I+3)} = \left\{\frac{1}{I+1} - \frac{1}{I+2}\right\} \cdot \frac{1}{I+3}$$

$$= \frac{1}{(I+1)(I+3)} - \frac{1}{(I+2)(I+3)}$$

$$= \frac{1}{2}\left\{\frac{1}{I+1} - \frac{1}{I+3}\right\} - \left\{\frac{1}{I+2} - \frac{1}{I+3}\right\}$$

$$= \left\{\frac{1}{2}L[|I+1|] + \frac{1}{2}L[|I+3|] - L[|I+2|]\right\}'.$$

Exercise 5. Evaluate the integral

$$\int_0^1 \frac{3I^2+1}{(I+1)^2(I+2)(I+3)}.$$

Note that

$$\frac{3I^2+1}{(I+1)^2(I+2)(I+3)} = \frac{3(I^2+2I+1) - 6(I+1) + 4}{(I+1)^2(I+2)(I+3)}$$

$$= \frac{3}{(I+2)(I+3)} - \frac{6}{(I+1)(I+2)(I+3)}$$

$$+ \frac{4}{(I+1)^2(I+2)(I+3)}.$$

Exercise 6. If $[a, b]$ is an interval not containing 0, -2, or 2, evaluate

$$\int_a^b \frac{1+I^2}{I(I^2-4)}.$$

Exercise 7. Show that $\int_a^1 \frac{1}{9I^2-16} = -\frac{L(7)}{24}.$

Examples. To obtain a simple graph whose derivative is

$$\frac{1}{(1+I)^2(1+I^2)},$$

consider

$$\frac{1}{(1+I)^2(1+I^2)} + \frac{aI+b}{1+I^2} = \frac{aI^3 + (2a+b)I^2 + (a+2b)I + (b+1)}{(1+I)^2(1+I^2)},$$

where each of a and b is a number to be determined so that $I^2 + 1$ shall be a factor of the numerator of the fraction in the right-hand member. We use long division:

$$\begin{array}{r}
aI + (2a+b) \\
I^2 + 1{\overline{\smash{\big)}\,aI^3 + (2a+b)I^2 + (a+2b)I + (b+1)}} \\
\underline{aI^3 \qquad\qquad\quad + \qquad aI} \\
(2a+b)I^2 + \qquad 2bI + (b+1) \\
\underline{(2a+b)I^2 \qquad\qquad + (2a+b)} \\
2bI + (1-2a)
\end{array}$$

The remainder is 0 if $b = 0$ and $a = \frac{1}{2}$. We then have

$$\frac{1}{(1+I^2)^2(1+I^2)} + \frac{\frac{1}{2}I}{1+I^2} = \frac{\frac{1}{2}I+1}{(1+I)^2}$$

or

$$\begin{aligned}
\frac{1}{(1+I)^2(1+I)^2} &= -\frac{1}{4}\cdot\frac{2I}{1+I^2} + \frac{1}{2}\cdot\frac{1}{1+I} + \frac{1}{2}\cdot\frac{1}{(1+I)^2} \\
&= \left\{-\frac{1}{4}L[1+I^2] + \frac{1}{2}L[|1+I|] - \frac{1}{2}\cdot\frac{1}{1+I}\right\}'.
\end{aligned}$$

Exercise 8. Show that

$$\frac{1}{I^2+I+1} = \frac{1}{I^2+I+\frac{1}{4}+\frac{3}{4}} = \frac{1}{\left(I+\frac{1}{2}\right)^2+\frac{3}{4}} = \left\{A\left[\frac{2I+1}{\sqrt{3}}\right]\right\}'\cdot\frac{2}{\sqrt{3}}.$$

Hence find the integral from 0 to 1 of this simple graph.

Exercise 9. Evaluate

(i) $\displaystyle\int_a^b \frac{1}{I^2+4}$,

(ii) $\displaystyle\int_a^b \frac{1}{(I^2+1)(I^2+I+1)}$,

(iii) $\displaystyle\int_0^2 \frac{1}{I^3+1}$.

A Theorem on Transformation of Integrals

Theorem. *Suppose that*

(i) $[a, b]$ is an interval, g is a simple graph whose X-projection is the interval $[p, q]$ such that

$$g(p) = a \qquad and \qquad g(q) = b,$$

the X-projection of g′ is [p, q], and g′ has property S at each of its points,

(ii) f is a simple graph whose X-projection is the Y-projection [r, s] of g and has property S at each of its points, and

(iii) h is the simple graph defined by

$$h(x) = \begin{cases} \int_a^x f, & a < x \leq s, \\ 0, & x = a, \\ -\int_x^a f, & r \leq x < a, \end{cases}$$

so that h′ = f. Then

$$\int_a^b f = \int_a^b h' = \int_p^q \{H[g]\}' = \int_p^q h'[g] \cdot g' = \int_p^q f[g] \cdot g';$$

i.e.,

$$\int_a^b f = \int_p^q f[g] \cdot g'.$$

(See Figure 7.2) This result is almost indispensable in evaluating certain integrals.

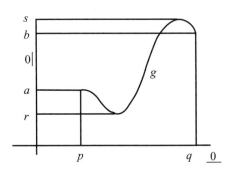

Figure 7.2

Examples.

$$\int_0^1 \frac{1}{(1+I^2)^2} = \int_0^{\pi/4} \frac{1}{(1+T^2)^2} \cdot T' = \int_0^{\pi/4} \frac{1}{(K^4)} \cdot K^2$$

$$= \int_0^{\pi/4} C^2 = \int_0^{\pi/4} \left\{ \frac{CS+I}{2} \right\}' = \frac{1}{4}\left(1 + \frac{\pi}{4}\right).$$

The area of the region $[Q[1 - I^2]; 0, 1]$, which is one-fourth the area of a circular disc of radius 1, is

$$\int_0^1 Q[1 - I^2] = \int_0^{\pi/2} Q[1 - S^2] \cdot S' = \int_0^{\pi/2} C^2 = \int_0^{\pi/2} \left\{ \frac{CS + I}{2} \right\}' = \frac{\pi}{4}.$$

Hence the area of a circular disc of radius 1 is π.

Exercise 10. If r is a positive number, find the area of one-fourth the circular disc of radius r,

$$\int_0^r Q[r^2 - I^2].$$

Exercise 11. If k is a number between 0 and 1,

$$\int_0^{\pi/2} \frac{1}{1 - kS} = \int_0^1 \frac{1}{1 - kS[2A]} \cdot 2A'.$$

Evaluate this integral.

Exercise 12. Evaluate the integral $\int_0^{\pi/2} \frac{C}{S + C + 1}.$

Exercise 13. Evaluate

(i) $\int_0^1 Q[4 + 9I^2],$

(ii) $\int_o^3 \frac{I}{Q[1 + I]},$

(iii) $\int_{-2}^{-1} Q[I^2 - 1],$ and

(iv) $\int_0^1 \frac{1}{2 + Q}.$

Problem. Try to invent a definition of *length* of a simple graph.

8

Computation Formulas Obtained by Means of the Integral

The approximation formulas for L and A were obtained by experimentation and conjecture. The integral furnishes a method for obtaining such formulas for many simple graphs.

We need a few preliminaries. First, it is convenient to make the agreement that

$$\int_a^a f = 0 \quad \text{and} \quad \int_b^a f = -\int_a^b f.$$

With this agreement, $L(x) = \int_1^x \frac{1}{I}$ for every positive number x and $A(x) = \int_0^x \Omega$ for every number x. Second, the following lemma, which we leave for the reader to prove, will be useful.

Lemma. *Suppose each of f and g is a simple graph with X-projection the interval $[a, b]$ having property S at each of its points and $g(x) \geq 0$ for every number x in $[a, b]$ or $g(x) \leq 0$ for every number x in $[a, b]$. Then there exists a number c in the interval $[a, b]$ such that*

$$\int_a^b fg = f(c) \cdot \int_a^b g.$$

Computation Formulas For L, A

We begin by rederiving the formulas for $L(c)$ found previously. If $c > 0$,

$$L(c) = \int_1^c H = \int_1^c H \cdot (I - k)',$$

where k is any number. Hence,

$$L(c) = \int_1^c \{H \cdot (I - k)\}' - \int_1^c (I - k) \cdot H' = \frac{c - k}{c} - (1 - k) + \int_1^c \frac{I - k}{I^2}.$$

The simplest formulas result if we take k to be 1 or c:

$$L(c) = \frac{c - 1}{c} + \int_1^c \frac{I - 1}{I^2}$$

and

$$L(c) = (c - 1) + \int_1^c \frac{I - c}{I^2}.$$

The first may be written

$$L(c) = \frac{c - 1}{c} + \int_1^c I \cdot \left\{ \frac{1}{2} \left(\frac{I - 1}{I} \right)^2 \right\}'.$$

so that if $c \neq 1$ it follows from the Lemma that

$$L(c) = \frac{c - 1}{c} + \frac{1}{2} \left(\frac{c - 1}{c} \right)^2 \cdot c',$$

where c' is in the interval $[c, 1]$ if $0 < c < 1$ and in the interval $[1, c]$ if $1 < c$.

To obtain a better approximation, we write

$$
\begin{aligned}
L(c) &= \frac{c - 1}{c} + \int_1^c \left(\frac{1}{I} \right)^2 \cdot \left\{ \frac{(I - 1)^2}{2} \right\}' \\
&= \frac{c - 1}{c} + \int_1^c \left\{ \left(\frac{1}{I} \right)^2 \cdot \frac{(I - 1)^2}{2} \right\}' - \int_1^c \frac{(I - 1)^2}{2} \cdot \left\{ \left(\frac{1}{I} \right)^2 \right\}' \\
&= \frac{c - 1}{c} + \frac{1}{2} \left(\frac{c - 1}{c} \right)^2 + \int_1^c I \cdot \left\{ \frac{1}{3} \left(\frac{I - 1}{I} \right)^3 \right\}' \\
&= \frac{c - 1}{c} + \frac{1}{2} \left(\frac{c - 1}{c} \right)^2 + \frac{1}{3} \left(\frac{c - 1}{c} \right)^3 \cdot c',
\end{aligned}
$$

where $c' = 1$, $c' = c$, or c' is a number between 1 and c.

Next, we write

$$L(c) = \frac{c-1}{c} + \frac{1}{2}\left(\frac{c-1}{c}\right)^2 + \int_1^c \left(\frac{1}{I}\right)^3 \cdot \left\{\frac{(I-1)^3}{3}\right\}'$$

$$= \frac{c-1}{c} + \frac{1}{2}\left(\frac{c-1}{c}\right)^2 + \frac{1}{3}\left(\frac{c-1}{c}\right)^3 + \int_1^c I \cdot \left\{\frac{1}{4}\left(\frac{I-1}{I}\right)^4\right\}'$$

$$= \frac{c-1}{c} + \frac{1}{2}\left(\frac{c-1}{c}\right)^2 + \frac{1}{3}\left(\frac{c-1}{c}\right)^3 + \frac{1}{4}\left(\frac{c-1}{c}\right)^4 \cdot c', 1 \gtreqless c' \gtreqless c,$$

and so on.

Exercise. Follow up the case $k = c$ to get

$$L(c) = (c-1) + \int_1^c \left(\frac{1}{I}\right)^2 \cdot \left\{\frac{(I-c)^2}{2}\right\}' = (c-1) - \frac{1}{2}(c-1)^2 \cdot \left(\frac{1}{c'}\right)^2,$$

where $c' = c, c' = 1$, or c' is a number between c and 1,

$$L(c) = (c-1) - \frac{1}{2}(c-1)^2 + \frac{1}{3}(c-1)^3 \left(\frac{1}{c'}\right)^3,$$

and so on.

Find the approximation formulas for A by this process, starting with

$$A(c) = \int_0^c \frac{1}{1+I^2} \cdot (I-k)'.$$

Computation Formulas for $E, \mathcal{S}, \mathcal{C}$

If c is a number,

$$E(c) = 1 + \int_0^c E' = 1 + \int_0^c E = 1 + \int_0^c E \cdot (I-k)',$$

where k is any number. Hence,

$$E(c) = 1 + \int_0^c \{E \cdot (I-k)\}' - \int_0^c (I-k) \cdot E'$$

$$= 1 + E(c) \cdot (c-k) - E(0) \cdot (-k) - \int_0^c (I-k) \cdot E.$$

Taking k to be c, we have

$$E(c) = 1 + c - \int_0^c E \cdot \left\{\frac{(I-c)^2}{2}\right\}' = 1 + c + \frac{c^2}{2} \cdot E(c'),$$

where c' is 0, c, or a number between 0 and c.

Exercise. Show that if c is a number and n is a positive integer, there exists a number c' such that $0 \leq c' \leq c$ or $c \leq c' \leq 0$ and

$$E(c) = 1 + \frac{c}{1} + \frac{c^2}{1 \cdot 2} + \cdots + \frac{c^n}{1 \cdot 2 \cdots \cdots n} + \frac{c^{n+1}}{1 \cdot 2 \cdots \cdots (n+1)} \cdot E(c').$$

If $n = 0$,

$$E(c) = 1 + c \cdot E(c').$$

Sufficiently many terms of the sum give, $E(1) = e = 2.718281828$, to nine decimal places. Show that, if c is a number and n is a positive integer, there exists a number c' such that $c \leq c' \leq 0$ or $0 \leq c' \leq c$ and

$$S(c) = c + \frac{c^3}{1 \cdot 2 \cdot 3} + \cdots + \frac{c^{2n-1}}{1 \cdot 2 \cdots \cdots (2n-1)} + \frac{c^{2n+1}}{1 \cdot 2 \cdots \cdots (2n+1)} \cdot C(c').$$

Obtain the analogous formula for C.

Exercise.

 (i) Show that if n is a positive integer and h is a positive number, there exists a vertical line $k|$ such that every point of the simple graph $\frac{E}{(I^n)}$ to the right of $k|$ is above $\underline{h}$.

 (ii) Show that if each of a and b is a positive number

$$(1 + a)(1 + b) < E(a + b)$$

and

$$\left(1 - \frac{a}{1+a}\right)\left(1 - \frac{b}{1+b}\right) > \frac{1}{E(a+b)}.$$

 (iii) Show that if c is a number, there exists a positive number d such that if x is a number distinct from 0 that differs from 0 by less than d, then $(1 + x)^{1/x}$ differs from e by less than c.

The Binomial Theorem

Theorem. *If m is a number and c is a number distinct from 0 and greater than -1 and n is a positive integer, there exists a number c' such that $c \leq c' \leq 0$ or*

$0 \le c' \le c$ *and*

$$(1 + c)^m = 1 + \int_0^c \{(1 + I)^m\}' = 1 + \frac{m}{1}c + \frac{m(m-1)}{1 \cdot 2}c^2 + \cdots$$
$$+ \frac{m(m-1)\cdots(m-n+1)}{1 \cdot 2 \cdots \cdot n}c^n$$
$$+ \frac{m(m-1)\cdots(m-n)}{1 \cdot 2 \cdots \cdot (n+1)}c^{n+1} \cdot \frac{(1+c')^{m+1}}{1+c}.$$

If $n = 0$,

$$(1 + c)^m = 1 + \frac{m}{1}c \cdot \frac{(1+c')^{m+1}}{1+c}.$$

Note that

$$0 < \frac{(1+c')^{m+1}}{1+c} \le \begin{cases} (1+c)^m & c > 0, m+1 > 0 \text{ or } c < 0, m+1 < 0, \\ \dfrac{1}{1+c} & c < 0, m+1 < 0 \text{ or } c < 0, m+1 > 0. \end{cases}$$

Example. To use this *binomial theorem* to compute approximations to $\sqrt{2}$, we first write 2 as a fraction such as $\frac{50}{25}$ or $\frac{98}{49}$ whose denominator is the square of a positive integer and whose numerator is *near* the square of a positive integer. Using $\frac{50}{25}$, we have

$$\sqrt{2} = \frac{1}{5}(7^2 + 1)^{1/2} = \frac{7}{5}\left(1 + \left[\frac{1}{7}\right]^2\right)^{1/2}$$
$$= \frac{7}{5}\left[1 + \frac{1}{2}\left(\frac{1}{7}\right)^2 - \frac{1}{8}\left(\frac{1}{7}\right)^4 + \frac{1}{16}\left(\frac{1}{7}\right)^6\right] + R,$$

where $|R| < .00000003$.

Exercise.

(i) Compute $2^{1/3}$ correct to six decimal places.

(ii) Suppose k is a positive number less than 1. Find approximation formulas for $\int_0^{\pi/2} (1 - k^2 S^2)^{1/2}$.

(iii) Show that if n is a positive integer

$$\left\{1 + \frac{1}{2} + \frac{1}{3} + \cdots + \frac{1}{n}\right\} - L(n+1) < \frac{5}{8}.$$

Length of a Simple Graph

The *inner product* of the point (x, y), denoted by P, and the point (u, v), denoted by Q, is the number $xu + yv$:

$$((P, Q)) = xu + yv.$$

We define the *sum* of P and Q, $P + Q$, and the *product* of k and P, $k \cdot P$, where k is a number by

$$P + Q = (x + u, y + v)$$

and

$$k \cdot P = (kx, ky).$$

Suppose each of P, Q, and R is a point. The inner product has the properties

(i) $((P, Q))$ is a number and $((P, P))$ is a positive number unless $P = (0, 0)$,

(ii) $((P + Q, R)) = ((P, R)) + ((Q, R))$,

(iii) $((P, Q)) = (Q, P))$,

and, if k is a number,

(iv) $((k \cdot P, Q)) = k((P, Q))$.

As a consequence of these properties we have

(v) $((P, Q))^2 \le ((P, P))((Q, Q))$.

If P is the point (x, y), the *negative* of P is $(-1) \cdot P$, so

$$-P = (-x, -y).$$

The *difference* $P - Q$ is the point $P + (-Q)$.

Definition. If P is a point and Q is a point, the *distance* from P to Q is the number

$$((P - Q, P - Q))^{1/2}$$

and is denoted by $|P - Q|$.

The distance from P to Q is a positive number unless $P = Q$ and the distance from P to Q is the distance from Q to P. The distance from (x, y) to $(0, 0)$ is $(x^2 + y^2)^{1/2}$, as previously defined. The distance from the point $(a, 0)$ to the

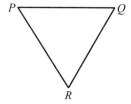

Figure 8.1

point $(b, 0)$ is, if $a < b$, the number $b - a$, which is the same as the length of the interval $[a, b]$. Distance has the property:

(vi) $|P - R| \leq |P - Q| + |Q - R|$,

which we will call the *triangle property* (see Figure 8.1).

We are now prepared to define the length of a simple graph.

Definition. The statement that the simple graph f *has length* on the interval $[a, b]$ means that the X-projection of f includes $[a, b]$ and there exists a number k such that, if D is a finite collection of nonoverlapping intervals filling up $[a, b]$, then the sum of all the distances

$$\left|(r, f(r)) - (s, f(s))\right|,$$

taken for all the intervals $[r, s]$ in D, does not exceed k. If f has length on $[a, b]$, the *least* such number k is called the *length* of f on $[a, b]$ and is denoted by

$$\ell_a^b f.$$

(See Figure 8.2.)

Theorem. *Suppose f is a simple graph whose X-projection includes the interval $[a, b]$. Then*

(i) *If f has length on $[a, b]$ and c is a number between a and b, then f has length on $[a, c]$ and on $[c, b]$ and*

$$\ell_a^b f = \ell_a^c f + \ell_c^b f.$$

(ii) *If the derivative f' of f is such that $Q\left[1 + (f')^2\right]$ is integrable on $[a, b]$, then*

$$\ell_a^b f = \int_a^b Q\left[1 + (f')^2\right].$$

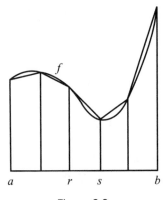

Figure 8.2

Exercise.

(i) Find the length on $[0, 1]$ of I^2.

(ii) Find the length on $[1, 2]$ of L.

(iii) Show that $\ell_0^{\pi/2} S = \sqrt{2} \int_0^{\pi/2} (1 - k^2 S^2)^{1/2}$ where $k = \frac{\sqrt{2}}{2}$ and find an approximation correct to four decimal places.

Theorem. *If f is a simple graph whose derivative f' has X-projection $[a, b]$ and property S at each of its points, and s is the simple graph defined by*

$$s(x) = \begin{cases} 0, & x = a, \\ \ell_a^x f, & a < x \leq b, \end{cases}$$

then, if $[u, v]$ is a subinterval of $[a, b]$,

$$\{s'(u) \cdot (v - u)\}^2 = \{v - u\}^2 + \{f'(u) \cdot (v - u)\}^2,$$

so that $s'(u) \cdot (v - u)$ is the length $[u, v]$ of the straight line tangent to f at the point $(u, f(u))$ (see Figure 8.3).

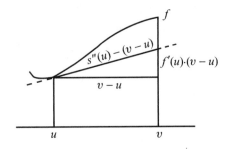

Figure 8.3

Length of an Arc of a Circle

The statement that M is the circle with center the point c and radius the positive number r means that M is the point set to which P belongs only if P is a point whose distance from c is r, or $|P - c| = r$. The circle with center $(0, 0)$ and radius r is the set of all points P such that $|P| = r$. The upper half of this circle is the simple graph $Q[r^2 - I^2]$, which we denote by q.

Exercise. Suppose x is a positive number less than r.

(i) q has length $r \cdot B(x/r)$ on $[0, x]$ and length $\ell_0^y q$ on $[x, r]$, where $y = q(x)$. (See Figure 8.4.)

(ii) The only positive number x less than r such that $x = q(x)$ is the number $\frac{r\sqrt{2}}{2}$, which we denote by c. Then

$$\ell_0^r q = \ell_0^c q + \ell_c^r q = \ell_0^c q + \ell_0^c q = 2r \cdot B\left(\frac{\sqrt{2}}{2}\right) = r \cdot \frac{\pi}{2}.$$

If $r = 1$, the length of the quarter circle is $\frac{\pi}{2}$.

(iii) Suppose the number θ is defined by

$$r \cdot \theta = \ell_x^r q.$$

Then

$$\theta = \frac{\pi}{2} - B\left(\frac{x}{r}\right) = \ell_{\frac{x}{r}}^1 Q[1 - I^2],$$

$$C(\theta) = \frac{x}{r} \quad \text{and} \quad S(\theta) = \frac{y}{r}.$$

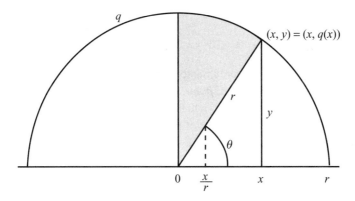

Figure 8.4

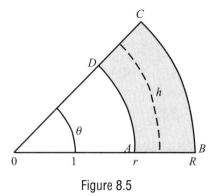

Figure 8.5

The number θ is the *argument* of the point (x, y), previously defined, and is the length on $[x/r, 1]$ of $Q[1 - I^2]$, $(x^2 + y^2 = r^2)$.

Exercise.

(i) The area of the shaded region in Figure 8.4 is $(1/2)\, r \cdot r \cdot (\pi/2 - \theta)$, or one-half the product of the radius of the circle times the length of q on $[0, x]$.

(ii) Suppose R is a number greater than r. Show that the area of the shaded region, $ABCD$, in Figure 8.5 is

$$\frac{R + r}{2} \cdot \theta \cdot (R - r),$$

or the product of the length of the arc h (see Figure 8.5) times the difference $R - r$ of the radii of the two circles.

Polar Equations

If (x, y) is a point, any ordered number pair (r, θ) such that $x = r\, C(\theta)$ and $y = r\, S(\theta)$ is said to be a pair of *polar coordinates* of the point (x, y). If f is a simple graph, the point set M to which (x, y) belongs only if there is a point (r, θ) of f such that (r, θ) is a pair of polar coordinates of (x, y) is called the *graph* of the polar equation $r = f(\theta)$.

Example. The circle with center $(1, 0)$ and radius 1 is the graph of the polar equation $r = 2C(\theta)$. (See Figure 8.6)

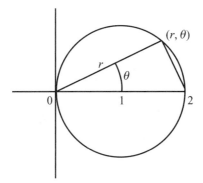

Figure 8.6

Exercise. Sketch the graph of

(i) $r = 4C(2\theta)$,

(ii) $r = 2\{1 - S(\theta)\}$,

(iii) $r = 1 + 2S(\theta)$.

The simple graph f shown in Figure 8.7 is the part of the graph of the polar equation $r = 2 \cdot C(2\theta)$ for which $0 \le \theta \le \frac{\pi}{4}$. We will find the area of the region $[f; 0, 2]$.

Let $[\theta_{i-1}, \theta_i]$, $i = 1, 2, \ldots, n$, be a finite collection of nonoverlapping intervals filling up $[0, \frac{\pi}{4}]$ and let $r_i = 2C(2\theta_i)$. The sum

$$\sum_{i=1}^{n} \frac{1}{2}r_i^2(\theta_i - \theta_{i-1}) = \sum_{i=1}^{n} 2C^2(2\theta_i)(\theta_i - \theta_{i-1})$$

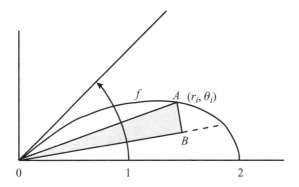

Figure 8.7

is the sum of the areas of n regions of which OAB in Figure 8.7 is typical. If the area of the region $[f; 0, 2]$ is the least number that no such sum exceeds, then the area is the integral

$$\int_0^{\pi/4} 2C^2[2I] = \int_0^{\pi/2} C^2 = \int_0^{\pi/2} \left\{ \frac{I + CS}{2} \right\}' = \frac{\pi}{4}.$$

Exercise. Find the area of the region bounded below by $\underline{0}$ and bounded above by the part of the graph of the polar equation $r = 2\{1 - C(\theta)\}$ for which $0 \le \theta \le \pi$.

Analogy Between $\mathcal{S}$, $\mathcal{C}$ and S, C

Suppose r is a positive number. The area u of the region OAB on the left in Figure 8.8 is

$$\frac{1}{2}xy + \int_x^r Q[r^2 - I^2].$$

If $\theta = 2u/(r^2)$, then $C(\theta) = \frac{x}{r}$ and $S(\theta) = \frac{y}{r}$. Now, the area u of the region OAB on the right in Figure 8.8 is

$$\frac{1}{2}xy - \int_r^x Q[I^2 - r^2].$$

If $\theta = 2u/(r)^2$ then $\mathcal{C}(\theta) = \frac{x}{r}$ and $\mathcal{S}(\theta) = \frac{y}{r}$.

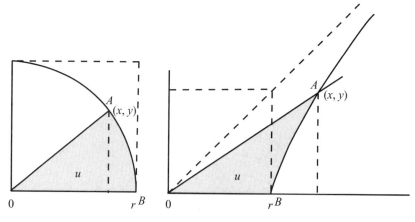

Figure 8.8

9

Simple Graphs Made to Order

We consider here the problem of constructing simple graphs, mainly as combinations of I, E, S, and C, that have certain prescribed properties.

The Equation $f' = g \cdot f + h$

If each of g and h is a simple graph whose X-projection is the set of all numbers having property S at each of its points and if (a, b) is a point, does there exist a simple graph f containing (a, b) such that $f' = g \cdot f + h$?

The simple graph E has the property that $E' = E$. Let us try to construct f from E:

$$f = vE[u].$$

Since $f' = vE[u]u' + v'E[u]$, we require that

(i) $vE[u]u' + v'E[u] = g \cdot vE[u] + h$

and

(ii) $v(a)E(u(a)) = b.$

Now, (i) is true if $u' = g$ and $v' = hE[-u]$ which, along with (ii), are true if

$$u(x) = \int_a^x g \quad \text{and} \quad v(x) = b + \int_a^x hE[-u].$$

Therefore, a simple graph f satisfying the given conditions is given by

$$f(x) = bE(u(x)) + \int_a^x hE[u(x) - u]$$

where

$$u(x) = \int_a^x g.$$

Suppose there is another such simple graph. Call it F, and let the difference be $w = F - f$. Then $w(a) = 0$ and $w' = gw$. The quotient $\frac{w}{E[u]}$ has slope 0 at each of its points and is therefore a horizontal line $\underline{k}$. So

$$w = kE[u].$$

Since $w(a) = 0$, we have $k = 0$ and so $w = 0$; and $F = f$. The problem has only one solution.

Exercise. Show that there are many simple graphs f such that $f(0) = 0$ and $f' = 2Q[f]$. One is $\underline{0}$. Find another.

The Equation $f'' + kf = g$

If k is a number, g is a simple graph whose X-projection is the set of all numbers having property S at each of its points, (a, b) is a point, and m is a number, does there exist a simple graph f containing (a, b) with slope m at this point such that $f'' + kf = g$?

Case 1. $k = 0$. There are two simple graphs α and β such that $\alpha'' = \underline{0}$, $\beta'' = \underline{0}$ and

$$\begin{aligned}
\alpha(a) &= 1, & \alpha'(a) &= 0, \\
\beta(a) &= 0, & \beta'(a) &= 1,
\end{aligned}$$

namely $\alpha = \underline{1}$ and $\beta = I - a$. We try to construct the solution f in the form

$$f = u\alpha + v\beta.$$

Since $f(a) = b$, we see it is necessary that

$$u(a) = b.$$

Differentiating, $f' = u\alpha' + \alpha u' + v\beta' + \beta v'$. Since f'' is involved, about the only way we can expect this to work is to have

$$\alpha u' + \beta v' = \underline{0},$$
$$f' = u\alpha' + v\beta', \qquad v(a) = m.$$

Then $f'' = u\alpha'' + v\beta'' + u'\alpha' + v'\beta' = v'$, so that we require

$$v' = g.$$

We have to determine u and v from the equations

$$u' + (I - a) \cdot v' = \underline{0},$$
$$v' = g,$$
$$u(a) = b, \quad v(a) = m.$$

Thus, $u' = -(I - a) \cdot g,\ v' = g,$

$$u(x) = b - \int_a^x (I - a) \cdot g, \quad v(x) = m + \int_a^x g,$$

and

$$f(x) = b + m \cdot (x - a) + \int_a^x (x - I) \cdot g.$$

This is the only simple graph with the prescribed properties.

Case 2. $k < 0$. We suppose $k = -r^2$ where r is a positive number. There are two simple graphs α and β such that $\alpha'' - r^2\alpha = \underline{0},\ \beta'' - r^2\beta = \underline{0}$ and

$$\alpha(a) = 1, \quad \alpha'(a) = 0,$$
$$\beta(a) = 0, \quad \beta'(a) = 1,$$

namely $\alpha = C[r(I - a)]$ and $r\beta = S[r(I - a)]$. If we try to construct the solution f in the form

$$f = u\alpha + v\beta$$

then we get

$$f(x) = bC(r(x - a)) + \frac{m}{r}S(r(x - a)) + \frac{1}{r}\int_a^x S[r(x - I)] \cdot g.$$

Case 3. $k > 0$. We suppose $k = r^2, r^2 > 0$. In this case, $\alpha = C[r(I - a)]$ and $r\beta = S[r(I - a)]$, we try to obtain f in the form $f = u\alpha + v\beta$, and we find that

$$f(x) = bC(r(x - a)) + \frac{m}{r}S(r(x - a)) + \frac{1}{r}\int_a^x S[r(x - I)] \cdot g.$$

The Equation $f'' + pf' + qf = g$

The problem of obtaining a simple graph f whose X-projection is the set of all numbers containing the point (a, b) and having slope m there such that

$f'' + pf' + qf = g$ can be reduced to the problems already discussed. It is understood that each of p and q is a number and g a simple graph whose X-projection is the set of all numbers having property S at each of its points. Trying the simplest things first, we try

$$f = uE[kI],$$

where k is a number. Then

$$f' = kuE[kI] + u'E[kI],$$
$$f'' = k^2uE[kI] + 2ku'E[kI] + u''E[kI],$$

and $f'' + pf' + qf = E[kI]\{u'' + (2k + p)u' + (k^2 + kp + q)u\}.$

If we choose k to be $-\frac{p}{2}$, our problem is reduced to the problem of finding a simple graph u for which

$$u'' + \frac{4q - p^2}{4}u = E\left[-\frac{p}{2}I\right] \cdot g.$$

Exercise. Find a simple graph f containing the point (a, b) with slope m at this point that satisfies:

(i) $f'' + 2f' + f = g,$

(ii) $f'' - 2f' + 2f = g,$

(iii) $f'' + 2f' + 10f = g,$

Here g is a simple graph with X-projection the set of all numbers having property S at each of its points.

10

More about Integrals

The integral is closely tied to the derivative. For example, to express a simple graph f as an integral by the formula $f(x) = f(a) + \int_a^x f'$ requires that f' be integrable on the interval over which the integral is extended. To find a more general kind of integral, we begin by generalizing the notion of length of an interval.

g-Length of an Interval

Definition. The statement that $g|_a^b$ is the g-*length* of the interval $[a, b]$ means that g is a simple graph whose X-projection includes $[a, b]$ and

$$g\Big|_a^b = g(b) - g(a).$$

In particular, the *I-length* of $[a, b]$ is $b - a$, the ordinary length of $[a, b]$.

Suppose that W is a metal wire of length c, that P_0 denotes one end of W, and that, if x is a positive number not greater than c, P_x denotes the place on W a distance x from P_0. Let g be the simple graph whose X-projection is $[0, c]$ defined by

$$g(x) = \begin{cases} 0, & x = 0, \\ \text{the mass of the piece of} & 0 < x \le c. \\ W \text{ from } P_0 \text{ to } P_x, \end{cases}$$

See Figure 10.1.

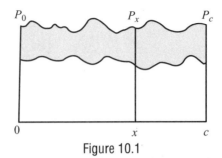

Figure 10.1

Then the g-length of the subinterval $[a, b]$ of $[0, c]$ is the mass of that piece of W from P_a to P_b.

As another illustration, let P_0 be a place above the surface of the earth and c the time for a falling body to pass from P_0 to the place P_c on the earth. For each number t in $[0, c]$, P_t denotes the place occupied by the body at time t. Let f denote the simple graph such that

$$f(t) = \begin{cases} 0, & t = 0, \\ \text{the distance from } P_0 \text{ to } P_t, & 0 < t \leq c. \end{cases}$$

The f-length of the subinterval $[a, b]$ of $[0, c]$ is the distance from P_a to P_b.

g-length has the fundamental property that, if c is a number between a and b, then

$$g\Big|_a^b = g\Big|_a^c + g\Big|_c^b.$$

If g' has X-projection $[a, b]$ the g-length of $[a, b]$ is expressible in terms of ordinary length by

$$g\Big|_a^b = g'(c) \cdot I\Big|_a^b$$

where c is some number between a and b.

The statement that the simple graph g is *nondecreasing* on the interval $[a, b]$ means that if $[p, q]$ is a subinterval of $[a, b]$, then $g\Big|_p^q \geq 0$. That is, the point of g with abscissa p is not higher than the point of g with abscissa q.

Integral of a Simple Graph with Respect to a Simple Graph

Suppose that the simple graph f is bounded on the interval $[a, b]$ and the simple graph g is nondecreasing on $[a, b]$.

The statement that $_iS$ is an inner sum for f with respect to g on $[a, b]$ means that there exists a finite collection D of nonoverlapping intervals filling up $[a, b]$ such that, if the g-length of each interval in D is multiplied by the largest number k which exceeds the ordinate of no point of f whose abscissa is in that interval, then the sum of all the products so formed is $_iS$. This is a generalization of inner sum for f with respect to I on $[a, b]$ used previously. The outer sum $_oS$ of f with respect to g on $[a, b]$ is the analogous generalization of outer sum used previously.

If V is the largest number which exceeds the ordinate of no point of f whose abscissa is in $[a, b]$ and if U is the smallest number exceeded by the ordinate of no point of f whose abscissa is in $[a, b]$, and if $_iS$ and $_oS$ are an inner and outer sum for f with respect to g on $[a, b]$, then

$$V \cdot g\Big|_a^b \leq {_iS} \leq {_oS} \leq U \cdot g\Big|_a^b.$$

This is trivial in case $_iS$ and $_oS$ are based on the *same* collection D of nonoverlapping intervals filling up $[a, b]$.

The smallest number that no inner sum for f with respect to g on $[a, b]$ exceeds is called the inner integral of f with respect to g on $[a, b]$ and is denoted by

$$_i\!\int_a^b f \, dg.$$

The largest number that exceeds no outer sum for f with respect to g on $[a, b]$ is called the outer integral of f with respect to g on $[a, b]$ and is denoted by

$$_o\!\int_a^b f \, dg.$$

In both these symbols, the letter d stands for the phrase "with respect to."

There is the following inequality

$$V \cdot g\Big|_a^b \leq {_iS} \leq {_i\!\int_a^b} f \, dg \leq {_o\!\int_a^b} f \, dg \leq {_oS} \leq U \cdot g\Big|_a^b.$$

Theorem. *If the simple graph f is bounded on the interval $[a, b]$ and the simple graph g is nondecreasing on $[a, b]$, then the following four statements are equivalent.*

(i) $_i\!\int_a^b f \, dg = {_o\!\int_a^b} f \, dg.$

(ii) *If c is a positive number, there exists an inner sum $_iS$ and an outer sum $_oS$ such that $_oS - {_iS} \leq c$.*

(iii) *If each of c and d is a positive number, there exists a finite collection D of nonoverlapping intervals filling up [a, b] such that, if there is an interval of D containing two numbers x and y such that f(x) and f(y) differ by more than c, then the sum of the g-lengths of all such intervals of D is less than d.*

(iv) *There exists a number J such that, if c is a positive number, there is a finite collection D of nonoverlapping intervals filling up [a, b] such that, if D' is a finite collection of nonoverlapping intervals filling up [a, b], with each end of each interval of D an end of an interval of D', and the g-length of each interval of D' is multiplied by the ordinate of any point of f whose abscissa is in that interval, then the sum of all the products so formed differs from J by less than c.*

Among these four mutually equivalent statements, these is one that does not make explicit use of the conditions imposed upon f and g. We accordingly take this statement, (iv), as our definition of integrability in the general case.

Definition. The statement that the simple graph f is integrable on $[a, b]$ with respect to the simple graph g means that each of f and g has X-projection including $[a, b]$ and that there exists a number J that satisfies the following: If c is a positive number, there is a finite collection D of nonoverlapping intervals filling up $[a, b]$ such that, if D' is a finite collection of nonoverlapping intervals filling up $[a, b]$, with each end of each interval of D an end of an interval of D', and the g-length of each interval in D' is multiplied by the ordinate of any point of f whose abscissa is in that interval, then the sum of all the products so formed differs from J by less than c. If f is integrable on $[a, b]$ with respect to g, the number J of the preceding statement is called the *integral* from a to b of f with respect to g and is denoted by

$$\int_a^b f\,dg.$$

The integral introduced in the earlier chapters is $\int_a^b f\,dI$. We shall continue to write $\int_a^b f$ for this integral.

Properties of $\int_a^b f\,dg$

It should be emphasized that if f is bounded and g nondecreasing on $[a, b]$, then any one of the four mutually equivalent statements in the preceding theorem may be taken as definition of the statement f is integrable with respect to g ("g-integrable") on $[a, b]$.

If the simple graph g has X-projection that includes the interval $[a, b]$ and if k is a number then $\underline{k}$ is g-integrable on $[a, b]$ and

$$\int_a^b \underline{k}\, dg = k \cdot g \mid_a^b.$$

In particular, if $a < x \le b$,

$$g(x) = g(a) + \int_a^b \underline{1}\, dg.$$

We agree that $\int_a^b f\, dg = 0$ (and $\int_b^a f\, dg = -\int_a^b f\, dg$). Then, $g(x) = g(a) + \int_a^x \underline{1}\, dg$ for every number x in $[a, b]$.

If f is g-integrable on $[a, b]$ and fg' is I-integrable on $[a, b]$, then

$$\int_a^b f\, dg = \int_a^b fg'.$$

In particular, if g' is I-integrable on $[a, b]$,

$$\int_a^b \underline{1}\, dg = g(b) - g(a) = \int_a^b g'.$$

If f is g-integrable on $[a, b]$, then g is f-integrable on $[a, b]$ and

$$\int_a^b f\, dg = (fg) \mid_a^b - \int_{a\cdot}^b g\, df.$$

If f is g-integrable on $[a, b]$ and k is a number, then

$$\int_a^b kf\, dg = \int_a^b f\, d(kg) = k \int_a^b f\, dg$$

and

$$\int_a^b f\, d(g + k) = \int_a^b f\, dg.$$

If each of f_1 and f_2 is g_1-integrable on $[a, b]$ and g_2-integrable on $[a, b]$, then

$$\int_a^b (f_1 + f_2)d(g_1 + g_2) = \int_a^b f_1\, dg_1 + \int_a^b f_1\, dg_2 + \int_a^b f_2\, dg_1 + \int_a^b f_2\, dg_2.$$

Theorem. *If f is g-integrable on $[a, b]$, then there exists a finite collection D of nonoverlapping intervals filling up $[a, b]$ such that, if $[p, q]$ is an interval* *of D on which f is unbounded, g(x) = g(p) for each number x in [p, q]. That is, g is constant on [p, q].*

We abbreviate the sum of products described in the definition of integrability as

$$\sum_{D'} f(x) \cdot g\Big|_p^q,$$

which is the sum of the g-length of [p, q] times the ordinate of the point of f whose abscissa is the number x of [p, q], taken for all [p, q] in D'.

Theorem. *Suppose each of f and g is a simple graph whose X-projection includes the interval [a, b]. The following two statements are equivalent:*

(i) *f is g-integrable on [a, b]*
    and

(ii) *if c is a positive number, there exists a finite collection D of nonoverlapping intervals filling up [a, b] such that, if D' is a finite collection of nonoverlapping intervals filling up [a, b] with each end of each interval of D an end of an interval of D', then*

$$\left| \sum_{D'} f(x) \cdot g\Big|_p^q - \sum_{D} f(y) \cdot \Big|_r^s \right| < c.$$

Corollary. *If f is g-integrable on [a, b] and if c is a number between a and b, then f is g-integrable on [a, c] and on [c, b] and $\int_a^b f\, dg = \int_a^c f\, dg + \int_c^b f\, dg$.*

Problem. Investigate conditions under which

$$\int_a^b f\, dF = \int_a^b fg\, dh$$

where

$$F(x) = \int_a^x g\, dh.$$

Bounded Variation

The statement that the simple graph g is of *bounded variation* on the interval [a, b] means that there exists a number k such that, if D is a finite collection of nonoverlapping intervals filling up [a, b], then the sum of the absolute values of the g-lengths of the intervals of D does not exceed k. If g is of bounded variation on $[a, b]$, the least such number k is denoted by $V_a^b g$ and is called the *total variation* of g on $[a, b]$.

If the simple graph g has X-projection including $[a, b]$ then the following two statements are equivalent:

(i) g has length on $[a, b]$,

and

(ii) g is of bounded variation on $[a, b]$.

Theorem. *If the simple graph f has X-projection $[a, b]$ and property S at each of its points and the simple graph g is of bounded variation on $[a, b]$, then f is g-integrable on $[a, b]$.*

Problem. If each of f and g is one of the simple graphs, $I, A, T, S, C,$ $K, B, L, E, S,$ or C, evaluate $\int_a^b f \, dg$.

Exercise. Suppose g is the simple graph defined by

$$g(x) = \begin{cases} -1/2, & x < 0, \\ 0, & x = 0, \\ 1/2, & x > 0. \end{cases}$$

If the X-projection of f includes $[-1, 1]$ and f has property S at $(0, f(0))$, then $\int_{-1}^1 f \, dg = f(0)$.

Theorem. *If the simple graph g is nondecreasing on $[a, b]$ and if f is bounded and g-integrable on $[a, b]$, then $|f|$ is g-integrable on $[a, b]$ and*

$$\left| \int_a^b f \, dg \right| \le \int_a^b |f| \, dg.$$

Moreover, if h is bounded and g-integrable on $[a, b]$, then fh is g-integrable on $[a, b]$ and

$$\left\{ \int_a^b fh \, dg \right\}^2 \le \left\{ \int_a^b f^2 \, dg \right\} \cdot \left\{ \int_a^b h^2 \, dg \right\}.$$

Extended Integrals

Theorem. *If f is a simple graph such that if h\ is a vertical line there is a point of f to the right of h\, then the following statements are equivalent:*

(i) *If c is a positive number, there exists a vertical line h\ such that if P and Q are points of f to the right of h\, then the ordinate of P differs from the ordinate of Q by less than c.*

(ii) *There exists a number k such that if α and β are horizontal lines with k between them, then there is a vertical line h\ such that every point of f to the right of h\ is between α and β.*

Example. If we add the point $(0, 1)$ to the simple graph S/I, the resulting simple graph has X-projection the set of all numbers and has property S at each of its points. Suppose this has been done, and we define the simple graph f by

$$f(x) = \int_0^x \frac{S}{I}.$$

If x and y are numbers and $0 < x < y$ then

$$f(y) - f(x) = \int_x^y \frac{S}{I} = \int_x^y \frac{1}{I} \cdot \{-C\}' = \frac{C(x)}{x} - \frac{C(y)}{y} - \int_x^y \frac{C}{(I^2)}$$

so that

$$|f(y) - f(x)| \le \frac{2}{x} + \int_x^y \left(-\frac{1}{I}\right)' = \frac{3}{x} - \frac{1}{y} < \frac{3}{x}.$$

Consequently, if c is a positive number and P and Q are points of f to the right of the vertical line $\frac{3}{c}|$, then the ordinate of P differs from the ordinate of Q by less than c, so (i) of the theorem is true. Hence (ii) is true. The number k of that statement we denote by

$$\int_0^{+\infty} \frac{S}{I},$$

which is called an *extended* integral.

We have seen that (ii) holds for the simple graph A with k the number $\frac{\pi}{2}$. Thus another extended integral is

$$\int_0^{+\infty} \frac{1}{1 + I^2} = \frac{\pi}{2}.$$

It is true that

$$\int_0^{+\infty} \frac{S}{I} = \frac{\pi}{2}.$$

Using this

$$\frac{1}{\pi} \int_0^{+\infty} \frac{S[xI]}{I} = \begin{cases} \frac{1}{2}, & x > 0, \\ 0, & x = 0, \\ -\frac{1}{2} & x < 0. \end{cases}$$

Exercise. Investigate in an analogous way the integrals

$$\int_0^x \frac{1 - C}{(I^2)} \quad \text{and} \quad \int_0^x E - I^2.$$

If the point $(0, \frac{1}{2})$ is added to $\frac{1-C}{I^2}$ the resulting simple graph has X-projection the set of all numbers and property S at each of its points. We suppose this has been done.

Number Sequences

The statement that s is a *number sequence* means that s is a simple graph whose X-projection is the set of all positive integers. If s is a number sequence, we will write s_n for $s(n)$, the ordinate of that point of s whose abscissa is the positive integer n. A number sequence s for which the equivalent statements (i) and (ii) of the last theorem are true is called a *convergent number sequence* and is said to *converge* to the *sequential limit* k, where k is the number in (ii).

Example. Suppose for each positive integer n, that e_n is one of the numbers 1 or -1 and

$$s_n = 1 + \frac{1}{2}e_1 + \left(\frac{1}{2}\right)^2 e_2 + \cdots + \left(\frac{1}{n}\right)^n e_n.$$

Then s is a convergent number sequence.

Suppose, for each positive integer n that a_n is a number, $s_n = a_1 + a_2 + \cdots + a_n$, and $t_n = |a_1| + |a_2| + \cdots + |a_n|$. Then

(i) The number sequence t is a convergent number sequence only in case there is a number h such that $t_n \leq h, n = 1, 2, 3, \ldots$.

(ii) If t is convergent, then s is convergent.

If $s_n = 1 - \frac{1}{2} + \frac{1}{3} - \cdots + (-1)^{n-1} \cdot \frac{1}{n}, n = 1, 2, 3, \ldots$, and $t_n = 1 + \frac{1}{2} + \frac{1}{3} + \cdots + \frac{1}{n}$, then s is convergent but t is not convergent.

Problem. Suppose for each positive integer n that x_n is a number, $x_n > x_{n+1} > 0$, the number sequence x has the sequential limit 0, and $s_n = x_n L(x_n)$. Show that s is a convergent number sequence with the sequential limit 0. Hence show that if the point $(0, 0)$ is added to the simple graph IL then the resulting simple graph has property S at $(0, 0)$.

Two Questions

(i) Investigate simple graphs having property Q defined as follows. The statement that the simple graph f has property Q at the point P means that P is a point of f and there exists a point P_R whose abscissa is the abscissa of P and a point P_L whose abscissa is the abscissa of P such that, if A_R and B_R are horizontal lines with P_R between them, and A_L and B_L are horizontal lines with P_L between them, then there exist vertical lines H and K with P between them such that every point of f between H and K to the right of P is between A_R and B_R and every point of f between H and K to the left of P is between A_L and B_L.

(ii) Investigate integrability defined as follows. The statement that f is *mean-integrable* with respect to g on $[a, b]$ means that each of f and g is a simple graph whose X-projection includes the interval $[a, b]$ and there exists a number J such that if c is a positive number, there is a finite collection D of nonoverlapping intervals filling up $[a, b]$ such that if D' is a finite collection of nonoverlapping intervals filling up $[a, b]$, with each end of each interval of D an end of some interval of D', and the g-length of each interval of D' is multiplied by one-half the sum of the ordinates of the points of f whose abscissas are the ends of that interval, then the sum of all the products so formed differs from J by less than c.

11

Simple Surfaces

The statement that f is a *simple surface* means that f is a collection, each element of which is an ordered pair (P, z), whose first member P is a point and whose second member z is a number such that no two ordered pairs in f have the same first member. The second member of that ordered pair in f whose first member is P is denoted by $f(P)$ (read f *of* P) or, if $P = (x, y)$, by $f(x, y)$ (read f *of x and y*).

Problem. Generalize to simple surfaces some of the ideas concerning simple graphs such as slope, property S, length, and integral.

To picture a simple surface, we regard the XY-plane as horizontal and represent the ordered pair (P, z) of f as a dot on the vertical line containing P at a distance z above the XY-plane if $z > 0$, at P if $z = 0$, and below the XY-plane a distance $|z|$ if $z < 0$ (see Figure 11.1).

The XY-projection of f is the point set to which P belongs only if P is the first member of an ordered pair in f. In Figure 11.1 the XY-projection of f is the rectangular interval $[ab; cd]$.

Gradient

To generalize to simple surfaces the notion of slope of a simple graph, we first reformulate our definition of slope of a simple graph. To avoid the use of pairs

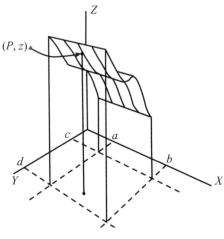

Figure 11.1

of vertical lines, we use the *segment*, which is the set of all numbers *between* two given numbers. That is, a segment is an interval without its ends.

Definition. The statement that the simple graph f has slope at the point $(x, f(x))$ of f means that there exists only one number m such that if c is a positive number then there exists a segment s containing x such that, if y is in s and in the X-projection of f, then

$$f(y) - f(x) = (y - x) \cdot m + |y - x| \cdot \left[\begin{array}{c} \text{a number between} \\ -c \text{ and } c \end{array} \right].$$

This is equivalent to the earlier definition. The requirement that there should be only one number m satisfying the condition implies that each two vertical lines with $(x, f(x))$ between them have between them a point of f distinct from $(x, f(x))$. The use of $|y - x|$ instead of $(y - x)$ amounts only to a possible change in sign in the number represented by the square-bracketed expression.

A *rectangular segment* is a rectangular interval without its edges.

Definition. The statement that the simple surface f has *gradient* at the ordered pair $((x, y), f(x, y))$ of f means that there exists only one ordered number pair $\{p, q\}$ such that if c is a positive number then there exists a rectangular segment s containing (x, y) such that, if (u, v) is in s and in the XY-projection of f, then

$$f(u, v) - f(x, y)$$
$$= (u - x) \cdot p + (v - y) \cdot q + |(u, v) - (x, y)| \cdot \left[\begin{array}{c} \text{a number between} \\ -c \text{ and } c \end{array} \right].$$

If f has gradient at $((x, y), f(x, y))$ then the ordered number pair $\{p, q\}$ is called the *gradient* of f at $((x, y), f(x, y))$.

The statement that the simple graph f is *continuous* at the point P means that f has property S at P. We have used the term *property S* instead of *continuous* because the word continuous could be misleading. Continuous, as applied to a simple graph, does not mean that the graph is all in one piece as the word might imply for physical things.

The statement that the simple surface f is continuous at $(P, f(P))$ means that $(P, f(P))$ is an ordered pair in f and if c is a positive number, there exists a rectangular segment s containing P such that if Q is a point of s belonging to the XY-projection of f, then $f(P)$ differs from $f(Q)$ by less than c.

Theorem. *If the simple surface f has gradient at $(P, f(P))$, then f is continuous at $(P, f(P))$.*

There are some interesting simple graphs connected with a simple surface f. Suppose h is a number that is the ordinate of a point in the XY-projection of f and denote by $f[I, h]$ the simple graph to which the point (x, y) belongs only if (x, h) is in the XY-projection of f and y is $f(x, h)$ (see Figure 11.2). Likewise, if k is a number that is the abscissa of a point in the XY-projection of f, then $f[k, I]$ denotes the simple graph to which the point (x, y) belongs only if (k, x) is in the XY-projection of f and y is $f(k, x)$.

If there is a number h such that the simple graph $f[I, h]$ has slope at one of its points, then f_1', called the 1-*derivative* of f, is the simple surface to which the ordered pair $((x, y), z)$ belongs only if $f[I, y]$ has slope at $(x, f[I, y](x))$ and $z = D_x f[I, y]$.

Example. If $f(x, y) = x^2 + 2xy + y^2$, then $f_1'(x, y) = D_x(I^2 + 2Iy + y^2) = 2x + 2y$.

Similarly, f_2', the 2-*derivative* of f, is the simple surface to which $((x, y), z)$ belongs only if $f[x, I]$ has slope at $(y, f[x, I](y))$ and $z = D_y f[x, I]$.

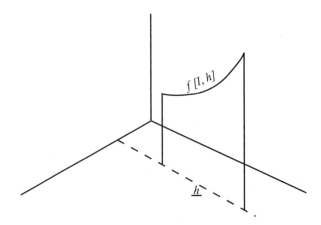

Figure 11.2

If f is the simple surface of the last example, then

$$f_2'(x, y) = D_y(x^2 + 2xI + I^2) = 2x + 2y.$$

For all appropriate points (x, y)

$$f_1'(x, y) = D_x f[I, y] \quad \text{and} \quad f_2'(x, y) = D_y f[x, I].$$

Exercise. Compute $f_1'(x, y)$ and $f_2'(x, y)$

(i) $f(x, y) = S(x) + C(y)$,

(ii) $f(x, y) = L(x^2 + 4y^2)$,

(iii) $f(x, y) = x^3 - 5x^2y + 2xy^2$.

The simple surfaces f_1' and f_2' are called the *first partial derivatives* of the simple surface f. There are four *second partial derivatives*, $(f_1')_1'$, $(f_1')_2'$, $(f_2')_1'$, and $(f_2')_2'$, denoted by f_{11}'', f_{12}'', f_{21}'', and f_{22}'', respectively, and called the 11-*derivative*, the 12-*derivative*, the 21-*derivative*, and the 22-*derivative* of f.

Theorem. *Suppose the simple surface f has XY-projection the rectangular interval $[ab; cd]$ and has gradient $\{p, q\}$ at $((x, y), f(x, y))$. Then*

$$p = f_1'(x, y) \quad \text{and} \quad q = f_2'(x, y).$$

Theorem. *If f, f_1', and f_2' have XY-projections including the rectangular segment s containing the point (x, y), f_1' is continuous at $((x, y), f_1'(x, y))$, and f_2' is continuous at $((x, y), f_2'(x, y))$, then f has gradient at $((x, y), f(x, y))$.*

Combination of Simple Surfaces

Suppose each of f and g is a simple surface and there is a point common to the XY-projection of f and the XY-projection of g. The *sum* $f + g$ is defined by

$$(f + g)(P) = f(P) + g(P)$$

and the product $f \cdot g$ by

$$(f \cdot g)(P) = f(P) \cdot g(P),$$

for every point P common to the XY-projection of f and the XY-projection of g; and, if P is a point such that $g(P) \neq 0$, the reciprocal $(1/g)$ is defined by

$$\frac{1}{g}(P) = \frac{1}{g(P)}$$

and the quotient (f/g) by $f \cdot (1/g)$.

Problem. Investigate continuity of the sum, product, and quotient of continuous f and g.

Theorem. *Suppose each of f and g is a simple surface and the rectangular segment s containing the point P is included in the common part of the XY-projections of f and g. If f has gradient at $(P, f(P))$ and g has gradient at $(P, g(P))$, then*

(i) $f + g$ has gradient $\{f_1'(P) + g_1'(P), f_2'(P) + g_2'(P)\}$ at $(P, (f + g)(P))$,

(ii) $f \cdot g$ has gradient

$$\{f(P)g_1'(P) + g(P)f_1'(P), f(P)g_2'(P) + g(P)f_2'(P)\}$$

at $(P, (f \cdot g)(P))$, and

(iii) if $g(P) \neq 0$, (f/g) has gradient

$$\left\{ \frac{g(P)f_1'(P) - f(P)g_1'(P)}{g^2(P)}, \frac{g(P)f_2'(P) - f(P)g_2'(P)}{g^2(P)} \right\} \; at \left(P, \frac{f}{g}(P) \right).$$

Suppose each of f, g, and h is a simple surface and there is a point P common to the XY-projection of g and the XY-projection of h and the point $(g(P), h(P))$ belongs to the XY-projection of f. The *bracket product* $f[g, h]$ (read f *of* g *and* h) is defined for all such P by

$$f[g, h](P) = f(g(P), h(P)).$$

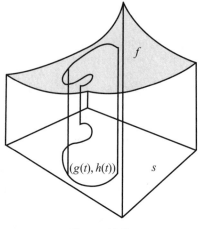

Figure 11.3

Under suitable conditions,

$$(f[g, h])'_1 = f'_1[g, h]g'_1 + f'_2[g, h]h'_1$$

and

$$(f[g, h])'_2 = f'_1[g, h]g'_2 + f'_2[g, h]h'_2.$$

More Simple Graphs Determined by a Simple Surface

Suppose each of g and h is a simple graph with X-projection the interval $[a, b]$ such that if t is in $[a, b]$, then $(g(t), h(t))$ is a point in the XY-projection of the simple surface f (see Figure 11.3). Another kind of bracket product is the simple graph $f[g, h]$ with X-projection $[a, b]$ defined for each t in $[a, b]$ by

$$f[g, h](t) = f(g(t), h(t)).$$

Theorem. *If the XY-projection of f is the rectangular segment s, f has gradient at each of its ordered pairs, and each of g' and h' has X-projection including the number t of $[a, b]$, then*

$$D_t f[g, h] = f'_1[g, h](t) \cdot g'(t) + f'_2[g, h](t) \cdot h'(t).$$

Example. If $g = x + uI$ and $h = y + vI$, where each of $x, y, u,$ and v is a suitable number, then

$$\begin{aligned} D_t f[g, h] &= D_t f[x + uI, y + vI] \\ &= f_1'(x + ut, y + vt) \cdot u + f_2'(x + ut, y + vt) \cdot v. \end{aligned}$$

Theorem. *Suppose f is a simple surface that has gradient at each of its ordered pairs and whose XY-projection is the rectangular segment s and g is a simple graph whose X-projection is the interval $[a, b]$, continuous at each of its points, and for each t in $[a, b]$ the point $(t, g(t))$ is in s and $f(t, g(t)) = 0$. Then for each number t in $[a, b]$ for which $f_2'(t, g(t)) \neq 0$,*

$$g'(t) = -\frac{f_1'(t, g(t))}{f_2'(t, g(t))}.$$

Example. If $f(x, y) = x^2 + y^2 - 1$ and g is one of the simple graphs $Q[1 - I^2]$ or $-Q[1 - I^2]$, then $f(t, g(t)) = 0$ for every number t in the interval $[-1, 1]$. Also, $f_1'(x, y) = 2x$ and $f_2'(x, y) = 2y$. So for $-1 < t < 1$

$$g'(t) = -\frac{t}{g(t)},$$

that is, $$D_t Q[1 - I^2] = -\frac{t}{Q(1 - t^2)}.$$

Exercise. If, for every point (x, y) such that $y > 0$, $f(x, y) = x - L(y)$, then $f(x, E(x)) = 0$ for every number x. Apply the preceding theorem to show that $E' = E$. In the same way, investigate $x - A(y)$ and $x - S(y)$.

g-Area of a Rectangular Interval

The area of the rectangular interval $[ab; cd]$ is the number $(b - a)(d - c)$ or $bd - bc - ad + ac$. If J is the simple surface with $J(x, y) = xy$ for every point (x, y), then the area of $[ab; cd]$ is $J(b, d) - J(b, c) - J(a, d) + J(a, c)$. This suggests a generalization of the notion of area.

Definition. The statement that $g|_a^b|_c^d$ is the *g-area* of the rectangular interval $[ab; cd]$ means that g is a simple surface whose XY-projection includes $[ab; cd]$ and

$$g\left.\left|_a^b\right|_c^d\right. = g(b, d) - g(b, c) - g(a, d) + g(a, c).$$

We note that

$$\begin{aligned}
g\left.\left|_a^b\right|_c^d\right. &= \{g[b, I] - g[a, I]\}\left|_c^d\right. \\
&= \{g(b, d) - g(a, d)\} - \{g(b, c) - g(a, c)\} = g\left.\left|_c^d\right|_a^b\right..
\end{aligned}$$

This generalization has the fundamental property of ordinary area that if D is a finite collection of rectangular intervals that are nonoverlapping and fill up

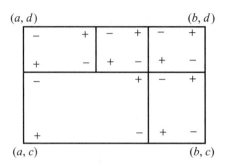

Figure 11.4

$[ab; cd]$ then the sum of the g-areas of the rectangular intervals of D is $g|_a^b|_c^d$. See Figure 11.4.

Theorem. *Suppose g is a simple surface whose XY-projection includes $[ab; cd]$ such that the XY-projection of the 12-derivative g''_{12} (or of the 21-derivatie g''_{21}) includes $[ab; cd]$. There exists a point (u, v) interior to $[ab; cd]$ (i.e., $a < u < b, c < v < d$) such that*

$$g\Big|_a^b\Big|_c^d = g''(u, v) \cdot (b - a)(d - c),$$

where g'' denotes g''_{12} or g''_{21} according to the case.

Corollary. *Suppose each of g, g''_{12}, and g''_{21} has XY-projection $[ab; cd]$ and each of g''_{12} and g''_{21} is continuous at each of its ordered pairs. Then $g''_{12} = g''_{21}$.*

If g is the simple surface defined by

$$g(x, y) = \begin{cases} 0, & (x, y) = (0, 0), \\ \dfrac{xy(x^2 - y^2)}{x^2 + y^2}, & (x, y) \neq (0, 0), \end{cases}$$

then $g''_{12}(0, 0) \neq g''_{21}(0, 0)$.

Maxima and Minima

Suppose f is a simple surface with XY-projection the rectangular segment s such that each of the second partial derivatives of f has XY-projection s and is continuous at each of its ordered pairs. Then $f''_{12} = f''_{21}$. If (x, y) and $(x + h, y + k)$ are points of s and, for each number t in $[0, 1]$, $F(t) = f(x + ht, y + kt)$, then

$$F(1) - F(0) = \int_0^1 F' = \int_0^1 F' \cdot (I - 1)' = F'(0) - \int_0^1 F'' \cdot \left[\frac{(I - 1)^2}{2}\right]'$$

so that there exists a number θ in the interval $[0, 1]$ such that

$$f(x+h, y+k) - f(x, y) = f_1'(x, y) \cdot h + f_2'(x, y) \cdot k$$

$$+ \frac{1}{2} \left\{ f_{11}''(x+\theta h, y+\theta k)h^2 \right.$$

$$+ 2f_{12}''(x+\theta h, y+\theta k)hk$$

$$\left. + f_{22}''(x+\theta h, y+\theta k)k^2 \right\}.$$

Corollary. *If* P *is a point of* s *such that* $f_1'(P) = 0$ *and* $f_2'(P) = 0$, *and* $f_{11}''(P)f_{22}''(P) - \{f_{12}''(P)\}^2 > 0$, *then*

(i) *if* $f_{11}''(P) > 0$, *there exists a rectangular segment* δ *containing* P *such that if* Q *is a point of* δ *distinct from* P *then* $f(Q) > f(P)$, *and*

(ii) *if* $f_{11}''(P) < 0$, *there exists a rectangular segment* δ *containing* P *such that if* Q *is a point of* δ *distinct from* P *then* $f(Q) < f(P)$.

Exercise. Let f be the simple surface defined for each point (x, y) such that $x + y \neq 0$ by

$$f(x, y) = \frac{xy(1 - xy)}{x + y}.$$

Find a point P such that for some rectangular segment δ containing P, $f(Q) < f(P)$ for every point Q in δ which is distinct from P.

Integral of a Simple Surface with Respect to a Simple Surface

The statement that the simple surface g *is nondecreasing* on $[ab; cd]$ means that the g-area of each rectangular interval included in $[ab; cd]$ is nonnegative. The statement that the simple surface f *is bounded* on $[ab; cd]$ means that there exists a number k such that, if P is in $[ab; cd]$ then $|f(P)| \leq k$.

If the simple surface f is bounded and g nondecreasing on $[ab; cd]$, inner sum, outer sum, inner integral, and outer integral are defined as with simple graphs. The statement that $_iS$ is an *inner sum* for f with respect to g on $[ab; cd]$ means that there exists a finite collection D of nonoverlapping rectangular intervals filling up $[ab; cd]$ such that if the g-area of each rectangular interval in D is multiplied by the largest number which does not exceed $f(P)$ for any point P in that rectangular interval, then the sum of all the products so formed is $_iS$. For *outer sum* $_oS$ replace, "largest number which does not exceed $f(P)$" by "smallest number which $f(P)$ does not exceed" in this statement. If

$V \leq f(P) \leq U$ for every point P in $[ab; cd]$, then

$$V \cdot g\big|_a^b\big|_c^d \leq {}_iS \leq {}_oS \leq U \cdot g\big|_a^b\big|_c^d.$$

The *inner integral* of f with respect to g on $[ab; cd]$, denoted by

$$\int_{i}{}_a^b \int_c^d f \, dg,$$

is the least number that no inner sum for f with respect to g on $[ab; cd]$ exceeds, and the *outer integral* of f with respect to g on $[ab; cd]$, denoted by

$$\int_{o}{}_a^b \int_c^b f \, dg,$$

is the largest number that exceeds no outer sum for f with respect to g on $[ab; cd]$. We have the inequality

$$V \cdot g\big|_a^b\big|_c^d \leq {}_iS \leq \int_{i}{}_a^b \int_c^d f \, dg \leq \int_{o}{}_a^b \int_c^d f \, dg \leq {}_oS \leq U \cdot g\big|_a^b\big|_c^d.$$

Theorem. *If f is bounded and g nondecreasing on $[ab; cd]$, then the following four statements are equivalent.*

(i) $\int_{i}{}_a^b \int_c^d f \, dg = \int_{o}{}_a^b \int_c^d f \, dg$.

(ii) If c is a positive number there exists an inner sum ${}_iS$ for f with respect to g on $[ab; cd]$ and an outer sum ${}_oS$ for f with respect to g on $[ab; cd]$ such that ${}_oS$ and ${}_iS$ differ by less than c.

(iii) If each of c and d is a positive number, there exists a finite collection D of nonoverlapping intervals filling up $[ab; cd]$ such that if there is a rectangular interval in D containing two points P and Q for which $f(P)$ differs from $f(Q)$ by more than c, then the sum of the g-areas of all such rectangular intervals in D is less than d.

(iv) There exists a number w such that if c is a positive number then there exists a finite collection D of nonoverlapping rectangular intervals filling up $[ab; cd]$ with the property that if D' is a finite collection of nonoverlapping rectangular intervals filling up $[ab; cd]$, with each point of each edge of a rectangular interval of D a point of an edge of some rectangular interval of D', and if the g-area of each rectangular interval in D' is multiplied by $f(P)$, where P is any point in that rectangular interval, then the sum of all the products so formed differs from w by less than c.

Definition. If each of f and g is a simple surface whose XY-projection includes the rectangular interval $[ab; cd]$ then the statement that f is *g-integrable* on $[ab; cd]$ means that there exists a number w for which (iv) of the last theorem is true. If f is g-integrable on $[ab; cd]$ then w is called the integral of f with respect to g on $[ab; cd]$ and is denoted by

$$\int_a^b \int_c^d f \, dg.$$

The statement that $\underline{\underline{k}}$ is a horizontal plane means that k is a number and $\underline{\underline{k}}$ the simple surface f such that for every point P, the second member of the ordered pair in f whose first member is P is the number k.

If the XY-projection of g includes $[ab; cd]$ and k is a number,

$$\int_a^b \int_c^d \underline{\underline{k}} \, dg = k \cdot g \Big|_a^b \Big|_c^d.$$

If f is g-integrable and fg'', where g'' is g_{12}'' or g_{21}'', is J-integrable on $[ab; cd]$, then

$$\int_a^b \int_c^d f \, dg = \int_a^b \int_c^d f g'' dJ.$$

$(J(x, y) = xy$ for every point (x, y).)

In particular if g'' is J-integrable on $[ab; cd]$ $(J(x, y) = xy)$, then

$$\int_a^b \int_c^d g'' \, dJ = g \Big|_a^b \Big|_c^d.$$

The statement that the simple surface g is of *bounded variation* on $[ab; cd]$ means the XY-projection of g includes $[ab; cd]$ and there exists a number k such that if D is a finite collection of nonoverlapping rectangular intervals filling up $[ab; cd]$ then the sum of the absolute values of the g-areas of the rectangular intervals of D does not exceed k.

Theorem. *If the simple surface f has XY-projection the rectangular interval $[ab; cd]$ and is continuous at each of its ordered pairs, and if the simple surface g is of bounded variation on $[ab; cd]$, then f is g-integrable on $[ab; cd]$.*

A sufficient condition for f to be continuous on $[ab; cd]$ is for f_1' and f_2' to be bounded on $[ab; cd]$.

Theorem. *Suppose the simple surface f has XY-projection the rectangular interval $[ab; cd]$ and, if c is a positive number, there exists a finite collection D of nonoverlapping rectangular intervals filling up $[ab; cd]$ such that if there is* *a rectangular interval in D containing a point P such that f is not continuous at $(P, f(P))$, then the sum of the areas of all such rectangular intervals is D is less than c. If f is bounded on $[ab; cd]$, then f is J-integrable on $[ab; cd]$.*

Example. If $g = J^2/4$, then $g'' = J$ and

$$\int_a^b \int_c^d J \, dJ = \frac{J^2}{4} \Big|_a^b \, \Big|_c^d = \frac{1}{4} \left\{ (ac)^2 - (bc)^2 + (bd)^2 - (ad)^2 \right\}.$$

Exercise. Evaluate

$$\int_a^b \int_c^d J^n \, dJ,$$

where n is an integer.

The statement that W is a *point in 3-space* means that W is an ordered number triple (x, y, z). Suppose f is a simple surface whose XY-projection includes the point set M. The region $[f; M]$ in 3-space determined by f and M is the point set in 3-space to which (x, y, z) belongs only if (x, y) is a point of M and z is 0, z is $f(x, y)$, or z is a number between 0 and $f(x, y)$.

If the XY-projection of f includes $[ab; cd]$ and $f(P) \geq 0$ for every point P in $[ab; cd]$, then the *volume* of the region $[f; [ab; cd]]$ in 3-space is the integral

$$\int_a^b \int_c^d f \, dJ.$$

Exercise. Find the volume of each of the regions in 3-space indicated:

(i) $[f; [01; 01]]$, where $f(x, y) = 12xy^2 + 8y$,

(ii) $[f; [12; 12]]$, where $f(x, y) = \dfrac{1}{(x + y)^2}$,

(iii) $[f; [01; 01]]$, where $f(x, y) = x \cdot E(x^2)$,

(iv) $[f; [0b; 0d]]$, where $b > 0, d > 0$ and $f(x, y) = Q(b^2 - x^2)$ (see Figure 11.5),

and

(v) $[f; [01; 01]]$, where $f(x, y) = 1 - x$.

Problem. Suppose r is a positive number and for each point (x, y) in $[0r; 0r]$

$$f(x, y) = \begin{cases} Q(r^2 - x^2 - y^2), & x^2 + y^2 \leq r^2, \\ 0, & x^2 + y^2 > r^2. \end{cases}$$

Sketch f and try to compute the volume of $[f; [0r; 0r]]$.

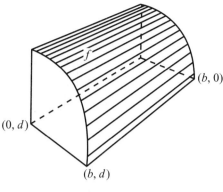

Figure 11.5

Iterated Integrals

Theorem. *Suppose the simple surface* f *has XY-projection* $[ab; cd]$ *and is continuous at each of its ordered pairs.*

(i) *If* h *is the simple graph with X-projection* $[a, b]$ *such that, for each number* x *in* $[a, b]$,

$$h(x) = \int_c^d f[x, I],$$

then h *is continuous at each of its points.*

(ii) *Suppose that for each point* (s, t) *in* $[ab; cd]$

$$h(s, t) = \int_0^t f[s, I]$$

and for each point (x, y) *in* $[ab; cd]$

$$g(x, y) = \int_a^x h[I, y].$$

Then $f = g_{12}''$.

Corollary. $\int_a^b \int_c^d f \, dJ = \int_a^b \int_c^d g_{12}'' \, dJ = g \, \big|_a^b \, \big|_c^d$,
or

$$\int_a^b \int_c^d f \, dJ = \int_a^b h[I, d],$$

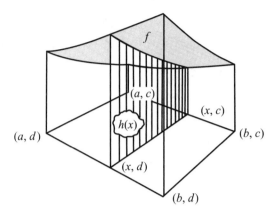

Figure 11.6

where

$$h(x, d) = \int_c^d f[x, I].$$

That is, with x fixed, $h(x, d)$ is the integral from c to d of $f[x, I]$ and $\int_a^b \int_c^d f \, dJ$ is the integral from a to b of $h[I, d]$.

It is convenient to write this as

$$\int_a^b \int_c^d f \, dJ = \int_a^b \left\{ \int_c^d f(x, y) \, dy \right\} dx.$$

Of course, there is also the formula

$$\int_a^b \int_c^d f \, dJ = \int_c^d \left\{ \int_a^b f(x, y) \, dx \right\} dy.$$

They are to be interpreted as in the corollary. The integrals in the right-hand members are called *iterated integrals*. In Figure 11.6, $h(x, d)$, or $h(x)$, is the area of a plane cross-section of the region $[f; [ab; cd]]$ in 3-space. The volume is then $\int_a^b h$.

Case Where f is not Continuous

Theorem.

(i) *If f is J-integrable on $[ab; cd]$, then f is bounded on $[ab; cd]$.*

(ii) If f is J-integrable on $[ab; cd]$, for each number x in $[a, b]$, $g(x) = {}_i\int_c^d f[x, I]$ and, for each number y in $[c, d]$, $h(y) = {}_i\int_a^b f[I, y]$, then

$$\int_a^b \int_c^d f \, dJ = \int_a^b g = \int_c^d h.$$

(iii) If $f[x, I]$ is integrable on $[c, d]$ for every x in $[a, b]$, then

$$\int_a^b \int_c^d f \, dJ = \int_a^b \left\{ \int_c^d f(x, y) \, dy \right\} dx$$

and, if $f[I, y]$ is integrable on $[a, b]$ for every y in $[c, d]$, then

$$\int_a^b \int_c^d f \, dJ = \int_c^d \left\{ \int_a^b f(x, y) \, dx \right\} dy.$$

Thus if f is J-integrable on $[ab; cd]$ and both the iterated integrals exist,

$$\int_a^b \int_c^d f \, dJ = \int_a^b \left\{ \int_c^d f(x, y) \, dy \right\} dx = \int_c^d \left\{ \int_a^b f(x, y) \, dx \right\} dy.$$

The iterated integrals may be written without the braces as

$$\int_a^b \int_c^d f(x, y) \, dy \, dx, \quad \text{and} \quad \int_c^d \int_a^b f(x, y) \, dx \, dy$$

or as

$$\int_a^b dx \int_c^d f(x, y) \, dy \quad \text{and} \quad \int_c^d dy \int_a^b f(x, y) \, dx.$$

Integral over a Point Set

Suppose M is a subset of the rectangular interval $[ab; cd]$. The statement that f is g-*integrable* on M means that each of f and g is a simple surface, the XY-projection of f includes M, that of g includes $[ab; cd]$ and, if F is the simple surface whose XY-projection is $[ab; cd]$ defined by

$$F(P) = \begin{cases} f(P), & P \text{ is in } M, \\ 0, & P \text{ is not in } M, \end{cases}$$

then F is g-integrable on $[ab; cd]$. The integral on M of f with respect to g is defined as the integral on $[ab; cd]$ of F with respect to g,

$$\iint_M f \, dg = \int_a^b \int_c^d F \, dg.$$

Theorem. *Suppose each of u and v is a simple graph with X-projection $[a, b]$ that is continuous at each of its points and $u(x) \leq v(x)$ for each number x in $[a, b]$. Suppose M is the point set to which (x, y) belongs only if x is in $[a, b]$ and y is $u(x)$ or y is $v(x)$, or y is a number between $u(x)$ and $v(x)$. If f is a simple surface whose XY-projection is M and that is continuous at each of its ordered pairs, then f is J-integrable on M and*

$$\iint_M f \, dJ = \int_a^b g,$$

where for each number x in $[a, b]$,

$$g(x) = \int_{u(x)}^{v(x)} f[x, I].$$

This may be written

$$\iint_M f \, dJ = \int_a^b \left\{ \int_{u(x)}^{v(x)} f(x, y) \, dy \right\} dx.$$

Exercise.

(i) Suppose M is the point set to which (x, y) belongs only if $x^2 + y^2 \leq 1$ and $f(x, y) = x^2 + y^2 + 1$. Evaluate $\iint_M f \, dJ$.

(ii) Suppose a is a positive number. Find the volume of the region in 3-space to which (x, y, z) belongs only if $x^2 + z^2 \leq a^2, x^2 + y^2 \leq a^2, x \geq 0, y \geq 0, z \geq 0$.

(iii) Suppose each of a, b, and c is a positive number. Find the volume of the region in 3-space to which (x, y, z) belongs only if

$$\left(\frac{x}{a}\right)^2 + \left(\frac{y}{b}\right)^2 + \left(\frac{z}{c}\right)^2 \leq 1, x \geq 0 \text{ and } z \geq 0. \text{ (Ans. } (\pi abc/6).)$$

Area of a Simple Surface

Addition, multiplication by numbers, and inner product are defined for points in 3-space as follows. If $P = (x, y, z)$, $Q = (u, v, w)$, and k is a number then $P + Q = (x + u, y + v, z + w), k \cdot P = (kx, ky, kz)$, and $((P, Q)) = xu + yv + zw$. We identify the point (x, y) in 2-space with the point $(x, y, 0)$ in 3-space. The distance from P to Q, denoted by $|P - Q|$, is the number $((P - Q, P - Q))^{1/2}$.

Definition. Suppose f is a simple surface with gradient $\{p, q\}$ at $(R, f(R))$. The statement that M is the *tangent plane to f at* $(R, f(R))$ means that M is the simple surface such that if P is the point (x, y) and $R = (a, b)$ then

$$M(P) = p \cdot (x - a) + q \cdot (y - b) + f(R).$$

Note that $M(R) = f(R)$, $M_1'(R) = p$, and $M_2'(R) = q$. If, for instance, R is contained in a rectangular segment included in the XY-projection of f, then

$$M_1'(R) = f_1'(R) \quad \text{and} \quad M_2'(R) = f_2'(R).$$

Definition. The statement that T is a *transformation* from the set A to the set B means that T is a collection of ordered pairs (a, b) such that a belongs to A, b to B, each element of A is the first member of an ordered pair in T, each element of B is the second member of an ordered pair in T, and no two ordered pairs in T have the same first member. If (a, b) is in T then b is denoted by $T(a)$ or by Ta.

Examples. A simple graph is a transformation from a number set to a number set and a simple surface is a transformation from a point set to a number set.

The statement that the transformation T from the set A to the set B is *reversible* means that no two ordered pairs in T have the same second member. That is, the collection U of ordered pairs to which (b, a) belongs only if (a, b) belongs to T is a transformation from the set B to the set A. U is called the *inverse* of the transformation T. For example, the simple graph L is a transformation from the set of all positive numbers to the set of all numbers. L is reversible, its inverse being the simple graph E that is a transformation from the set of all numbers to the set of all positive numbers.

Problem. Suppose f is a simple surface whose XY-projection includes the rectangular interval $[a, a + h; b, b + k]$, where h is a positive number and k a positive number. Suppose that f has gradient $\{p, q\}$ at $((a, b), f(a, b))$. Suppose M is the tangent plane to f at $((a, b), f(a, b))$,

$$M(x, y) = (x - a) \cdot p + (y - b) \cdot q + f(a, b),$$

and M_0 the subset of M whose XY-projection is $[a, a + h; b, b + k]$ (see Figure 11.7). What shall we mean by the *area* of M_0?

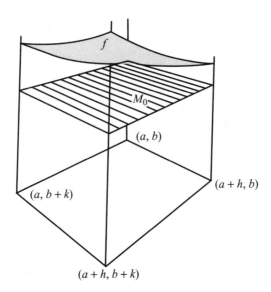

Figure 11.7

Let us identify the ordered pair $((x, y), M(x, y))$ of M with the point in 3-space $(x, y, M(x, y))$ and try to find a reversible transformation F from M_0 to a point set K such that distances are preserved under the transformation. That is, if $W_1 = (x, y, M(x, y))$ and $W_2 = (u, v, M(u, v))$, then

$$|W_1 - W_2| = |F(W_1) - F(W_2)|.$$

Such a transformation could be described as a rigid motion taking M_0 onto a region K in the XY-plane. If we can compute the area of K, we may assign to M_0 the same area.

For each point (x, y) we denote $(x - a, y - b)$ by (s, t), so that

$$x = a + s, \quad y = b + t.$$

Then

$$M(x, y) = f(a, b) + p \cdot s + q \cdot t$$

and

$$
\begin{aligned}
(x, y, M(x, y)) &= (a + s, b + t, f(a, b) + p s + q t) \\
&= (a, b, f(a, b)) + s \cdot (1, 0, p) + t \cdot (0, 1, q).
\end{aligned}
$$

Hence, if $W_0 = (a, b, f(a, b))$, $U_0 = (1, 0, p)$, $V_0 = (0, 1, q)$, then the point W in 3-space belongs to M only if there exists an ordered pair (s, t) such that

$$W = W_0 + s \cdot U_0 + t \cdot V_0.$$

This defines a reversible transformation from M to the set of all points, consisting of the ordered pairs $(W, (s, t))$.

If $W_1 = W_0 + s_1 \cdot U_0 + t_1 \cdot V_0$ and $W_2 = W_0 + s_2 \cdot U_0 + t_2 \cdot V_0$, the distance $|W_1 - W_2|$ is $((W_1 - W_2, W_1 - W_2))^{1/2}$ or

$$|W_1 - W_2| = \left\{ (s_1 - s_2)^2 |U_0|^2 + 2(s_1 - s_2)(t_1 - t_2)((U_0, V_0)) + (t_1 - t_2)^2 |V_0|^2 \right\}^{1/2}.$$

If $|U_0| = 1$, $((U_0, V_0)) = 0$, and $|V_0| = 1$, this is the distance from (s_1, t_1) to (s_2, t_2).

To obtain a distance-preserving transformation, suppose

$$U = \frac{1}{|U_0|} \cdot U_0$$

so that $|U| = 1$ and

$$U_0 = |U_0| \cdot U.$$

If c is a number then $((U, c \cdot U + V_0)) = c + ((U, V_0)) = 0$ provided $c = -((U, V_0))$. Then, if

$$V = \frac{1}{|V_0 - ((U, V_0)) \cdot U|} \cdot \{V_0 - ((U, V_0)) \cdot U\}$$

we have $|V| = 1$ and $((U, V)) = 0$. Also,

$$V_0 = |V_0 - ((U, V_0)) \cdot U| \cdot V + ((U, V_0)) \cdot U.$$

So the point W in 3-space belongs to M only if there exists an ordered number pair (s, t) such that

$$\begin{aligned} W &= W_0 + s \cdot |U_0| \cdot U + t \cdot \{|V_0 - ((U, V_0)) \cdot U| \cdot V + ((U.V_0)) \cdot U\} \\ &= W_0 + u \cdot U + v \cdot V \end{aligned}$$

where

$$u = s|U_0| + t((U, V_0)) \quad \text{and} \quad v = t|V_0 - ((U, V_0)) \cdot U|.$$

The relationship

$$W = W_0 + u \cdot U + v \cdot V$$

defines a reversible transformation F from M to the set of all points (u, v).

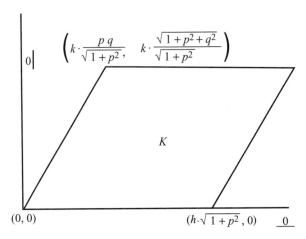

Figure 11.8

Since F is distance preserving, we have justified the name *plane* for M. There is a rigid motion which "carries" M onto the plane of all points (u, v).

The point (x, y) belongs to the XY-projection of M_0 only if (s, t) belongs to $[0, h; 0, k]$ and only if (u, v) belongs to the parallelogram disc shown in Figure 11.8. The area of this disc is

$$h \cdot k \cdot \sqrt{1 + p^2 + q^2},$$

which we take to be the area of M_0.

We can now make the following definition.

Definition. If f is a simple surface whose XY-projection includes the point set G, and if

$$Q\left[1 + (f_1')^2 + (f_2')^2\right]$$

is J-integrable on G, then *the area of f on G* is

$$\iint_G Q\left[1 + (f_1')^2 + (f_2')^2\right] dJ.$$

Exercise. Suppose that a is a positive number and M is the point set to which (x, y) belongs only if $4x^2 + y^2 \le a^2, x \ge 0$, and $y \ge 0$. If f is the simple surface whose XY-projection is M such that

$$f(x, y) = Q(a^2 - x^2 - y^2),$$

show that the area of f is $\pi a^2 / 6$.

Path Integrals

We denote by $\{p(P), q(P)\}$ the gradient of the simple surface f at $(P, f(P))$ and define the number $\phi(P, Q)$ for every point Q in the XY-projection of f by $\phi(P, P) = 0$ and, if $P = (x, y)$, $Q = (u, v)$, and $P \neq Q$, then

$$f(Q) - f(P) = (u - x) \cdot p(P) + (v - y) \cdot q(P) + |P - Q| \cdot \phi(P, Q).$$

If c is a positive number, there exists a positive number d such that if Q is in the XY-projection of f and $|P - Q| < d$, then

$$|\phi(P, Q)| < c.$$

If f has gradient at $(P, f(P))$ and at $(Q, f(Q))$, then

$$|\phi(P, Q)| \leq |p(Q) - p(P)| + |q(Q) - q(P)| + |\phi(Q, P)|.$$

The statement that the point set M is a *path* means that there exists an ordered pair (x, y), each element of which is a simple graph with X-projection the interval $[0, 1]$, continuous at each of its points, and having length, such that P belongs to M only if there exists a number t in $[0, 1]$ for which $P = (x(t), y(t))$. The path M is denoted by $\{x, y\}_0^1$ and the path is *closed* in case $x(0) = x(1)$ and $y(0) = y(1)$.

Theorem. *Suppose f is a simple surface and M the path $\{x, y\}_0^1$ and*

(i) if P is in M, f has gradient $\{p(P), q(P)\}$ at $(P, f(P))$;

(ii) if P is in M, p is continuous at $(P, p(P))$ and q at $(P, q(P))$; and

(iii) each of x' and y' has X-projection $[0, 1]$ and is continuous at each of its points.

Then,

$$\int_0^1 p[x, y]\, dx + \int_0^1 q[x, y]\, dy = \int_0^1 1\, df[x, y] = f[x, y]\Big|_0^1.$$

The same conclusion is true without the restrictive condition (iii) on x' and y'.

Theorem. *Suppose each of p and q is a simple surface and, if P is a point in the rectangular segment s, p is continuous at $(P, p(P))$ and q at $(P, q(P))$ and, if $\{x, y\}_0^1$ is a closed path that is a subset of s,*

$$\int_0^1 p[x, y]\, dx + \int_0^1 q[x, y]\, dy = 0.$$

Then there exists a simple surface f with gradient $\{p(P), q(P)\}$ at $(P, f(P))$ for each point P of s.

Vectors

A vector in the plane is an ordered number pair $\{A, B\}$, where A is the *horizontal component*, B is the *vertical component* and $Q(A^2 + B^2)$ is the *magnitude* of the vector $\{A, B\}$. The statement that θ is the *inclination* of the vector $\{A, B\}$ means that the magnitude r of $\{A, B\}$ is positive and θ is the number such that $0 \leq \theta < 2\pi$, $A = r \cdot C(\theta)$, and $B = r \cdot S(\theta)$. If P is the point (x, y) and Q the point (u, v), then the *vector from P to Q* is $\{u - x, v - y\}$.

Suppose f is a simple surface with gradient $\{p, q\}$ at $(P, f(P))$ and v is a vector with inclination θ. The *directional derivative* of f at $(P, f(P))$ *in the direction* v, denoted by

$$\frac{\partial f}{\partial v},$$

is the number $p \cdot C(\theta) + q \cdot S(\theta)$.

Theorem. *Suppose f is a simple surface with gradient $\{p, q\}$ at $(P, f(P))$ and $p^2 + q^2 > 0$. Then $\{p, q\}$ is a vector whose magnitude r is the maximum of the absolute values of the directional derivatives of f at $(P, f(P))$, $\partial f / \partial v = r$ if v is one of the vectors $\{p, q\}$ or $\{-p, -q\}$, and $\partial f / \partial v = -r$ if v is the other one of these vectors. The absolute value of the directional derivative of f is less than r in any other direction.*

Assorted Theorems and Problems

Theorems.

1. *Suppose f is a simple surface with gradient at each of its ordered pairs and XY-projection the rectangular segment s. If P is a point of s such that $f(P) = 0$, $f_2'(P) \neq 0$, and f_2' is continuous at $(P, f_2'(P))$, then there exists a segment t containing the abscissa of P and a simple graph g whose X-projection is t such that g is continuous at each of its points and, if x is in the segment t, $f(x, g(x)) = 0$. Moreover, g has slope $-f_1'(x, g(x))/f_2'(x, g(x))$ at $(x, g(x))$ for each x in t.*

2. *If the simple surface f has XY-projection $[ab; cd]$ and is continuous at each of its ordered pairs, if $f(P) \geq 0$ for every P in $[ab; cd]$, and if $\int_a^b \int_c^d f \, dJ = 0$, then $f(P) = 0$ for every P in $[ab; cd]$.*

3. If each of f, g, and h is a simple surface, h is nondecreasing on $[ab; cd]$, and each of f and g is bounded and h-integrable on $[ab; cd]$, then

$$\left(\int_a^b \int_c^d fg \, dh \right)^2 \le \left(\int_a^b \int_c^d f^2 \, dh \right) \left(\int_a^b \int_c^d g^2 \, dh \right).$$

Problems.

(i) Denote the number $\int_0^{+\infty} \frac{S}{I}$ by k. To prove that $k = \frac{\pi}{2}$, assume that $|k - (\pi/2)| = \delta > 0$, and try to reach a contradiction. Let f be the simple surface defined by $f(x, y) = E(-xy)S(y)$. If $a > 0$, and $b > 0$, then

$$\int_0^a \int_0^b f \, dJ = \int_0^a \left\{ \int_0^b E(-xy)S(y) \, dy \right\} dx$$

$$= \int_0^b \left\{ \int_0^a E(-xy)S(y) \, dx \right\} dy,$$

so that

$$\int_0^a \frac{1 - E[-bI]\{C(b) + I S(b)\}}{1 + I^2} = \int_0^b \left\{ 1 - E[-aI] \right\} \frac{S}{I}$$

or

$$\int_0^b \frac{S}{I} - A(a) = \int_0^b E[-aI] \frac{S}{I} - \int_0^a \frac{E[-bI]\{C(b) + I S(b)\}}{1 + I^2}.$$

(ii) Devise a way to compute approximations to

$$\int_0^b \frac{S}{I}$$

for any positive number b. A suggestion: Start with

$$\int_0^b \frac{S}{I} = \frac{\pi}{2} - C(b) \cdot \int_0^{+\infty} \frac{bE[-I]}{b^2 + I^2} - S(b) \cdot \int_0^{+\infty} \frac{I E[-I]}{b^2 + I^2},$$

a consequence of the formula in (i).

(iii) Show that

$$\int_0^{+\infty} E[-I^2] = \frac{1}{2} \sqrt{\pi}.$$

Denote this extended integral by k_0 and suppose that for each positive integer n

$$k_n = \int_0^{+\infty} I^n E[-I^2].$$

There is the relation $k_n = \frac{n-1}{2} \cdot k_{n-2}, n = 2, 3, 4 \ldots, k_1 = \frac{1}{2}$. Also $r_n = \int_0^{\pi/2} S^n$ can be computed, $n = 0, 1, 2 \ldots$. Note that $k_n k_{n+2} - k_{n+1}^2 > 0$, $n = 0, 1, 2 \ldots$, because $k_{n+2}t^2 - 2k_{n+1}t + k_n > 0$ for every number t.

Theorems.

1. *If the simple graph f has X-projection the interval $[a, b]$ and is of bounded variation on $[a, b]$, then there exists a simple graph g nondecreasing on $[a, b]$ and a simple graph h nondecreasing on $[a, b]$ such that*

$$f = g - h.$$

2. *If the simple surface f has XY-projection the rectangular interval $[ab; cd]$ and is of bounded variation on $[ab; cd]$, then there exists a simple surface g nondecreasing on $[ab; cd]$ and a simple surface h nondecreasing on $[ab; cd]$ such that*

$$f = g - h.$$

3. *If the simple graph f is nondecreasing on the interval $[a, b]$ and P is a point of f with abscissa in $[a, b]$, then f has property Q at the point P.*

4. *If the simple graph f is of bounded variation on $[a, b]$ and P is a point of f with abscissa in $[a, b]$, then f has property Q at P.*

5. *If the simple graph f has X-projection $[a, b]$ and has property Q at each of its points, then between any two numbers x and y of $[a, b]$ there is a number z such that f is continuous at $(z, f(z))$. If f is not continuous at one of its points and M is the number set to which x belongs only if x is the abscissa of a point of f where f is not continuous, then there exists for each positive integer n a number set G_n containing only one number of M and such that each number of M is in G_n for some positive integer n.*

6. *Suppose G is a collection of segments such that each number of the interval $[a, b]$ belongs to some segment of G. Then there exists a collection H of segments such that (1) H is a finite collection, (2) each segment of H is a segment of G, and (3) each number of $[a, b]$ belongs to some segment of H.*

7. *Suppose M is an infinite subset of the interval $[a, b]$. There exists a number c such that each segment containing c includes an infinite subset of M.*

8. *Suppose for each positive integer n, G_n is a number set such that if $[a, b]$ is an interval then there exists a subinterval of $[a, b]$ containing no number of G_n. If G is the set to which x belongs only if there is a positive integer n such that x belongs to G_n, then G does not fill up any interval.*

9. *Suppose f is a simple graph whose X-projection is the interval $[a, b]$, and if c is a positive number then there exists a simple graph g whose X-projection is $[a, b]$, continuous at each of its points, such that $f(x)$ differs from $g(x)$ by less than c for every x in $[a, b]$. Then f is continuous at each of its points.*

12

Successive Approximations

Existence of Certain Numbers Established by Successive Approximations

Suppose a is a positive number and we wish to prove the existence of a positive number x whose square is a. If x_0 is any positive number, then $x_0^2 > a$, $x_0^2 = a$, or $x_0^2 < a$. In the first case, x_0 is too large and

$$x_0 > \frac{a}{x_0}.$$

The average $(1/2)\{x_0 + (a/x_0)\}$ is less than x_0 and greater than a/x_0 and may be a better approximation to the hypothetical number x whose square is a than either x_0 or a/x_0. The same is possibly true in case $x_0^2 < a$. Thus, the number x_1, defined by

$$x_1 = \frac{1}{2}\left\{x_0 + \frac{a}{x_0}\right\},$$

may be a better approximation to x than x_0 is. Then x_2, defined by

$$x_2 = \frac{1}{2}\left\{x_1 + \frac{a}{x_1}\right\},$$

may be a still better approximation,

$$x_3 = \frac{1}{2}\left\{x_2 + \frac{a}{x_2}\right\}$$

still better, and so on.

To obtain experimental evidence in support of this conjecture, suppose a is 2 and $x_0 = 1.5$. Then

$$x_1 = \frac{1}{2}\left\{1.5 + \frac{2}{1.5}\right\} = 1.4167.$$

Since $x_0^2 = 2.25$ and $x_1^2 = 2.007$, the conjecture is supported. Then

$$x_2 = \frac{1}{2}\left\{1.4167 + \frac{2}{1.4167}\right\} = 1.414215$$

and $x_2^2 = 2.00004$. It appears that $x_1 > x_2$.

Lemma. *If a is a positive number, x_0 is a positive number, and, for each positive integer n,*

$$x_n = \frac{1}{2}\left\{x_{n-1} + \frac{a}{x_{n-1}}\right\},$$

then $x_n \geq x_{n+1}$.

In fact,

$$x_n - x_{n+1} = x_n - \frac{1}{2}\left\{x_n + \frac{a}{x_n}\right\} = \frac{1}{2}\left\{x_n - \frac{a}{x_n}\right\} = \frac{a}{8x_n}\left\{x_{n-1} - \frac{a}{x_{n-1}}\right\}^2,$$

so that $x_n - x_{n+1} \geq 0$ or $x_n \geq x_{n+1}$.

Since $x_n > 0$ for every positive integer n, there is a largest number r that does not exceed x_n for any positive integer n. The number r is nonnegative and if c is a positive number then there exists a number N such that x_n differs from r by less than c if $n > N$.

Now, if $\varepsilon_n = x_n - r$ or $x_n = r + \varepsilon_n$, then $r^2 - a = \varepsilon_n^2 - 2r\varepsilon_{n+1} - 2\varepsilon_n\varepsilon_{n+1}$. If we suppose $|r^2 - a| = c > 0$, there exists a number N such that $|\varepsilon_n^2 - 2r\varepsilon_{n+1} - 2\varepsilon_n\varepsilon_{n+1}| < c$, if $n > N$. So $c < c$. This contradiction shows that $r^2 = a$.

We have proved by successive approximations that there is a positive number whose square is a.

Our proof shows that for each positive integer n

$$\varepsilon_n^2 - 2r\varepsilon_{n+1} - 2\varepsilon_n\varepsilon_{n+1} = 0,$$

so

$$\varepsilon_{n+1} = \frac{\varepsilon_n^2}{2(r + \varepsilon_n)} < \frac{\varepsilon_n^2}{2r}.$$

Thus, x_{n+1} approximates r with an error about the square of the error in the approximation of x_n to r. So, if x_n is accurate to k decimal places, then x_{n+1} is accurate to about $2k$ decimal places.

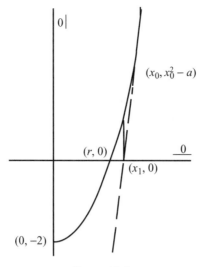

Figure 12.1

The problem just considered may be interpreted in the following way. Suppose f is the simple graph such that if x is a number then $f(x) = x^2 - a$ (see Figure 12.1 where $a = 2$). Since f is continuous and increasing for nonnegative abscissas, $f(0) < 0$, and $f(a + 1) > 0$, there exists only one number r such that

$$r > 0 \quad \text{and} \quad f(r) = 0.$$

Now if x_0 is a positive number, the tangent line to f at the point $(x_0, f(x_0))$ is

$$2x_0(I - x_0) + x_0^2 - a$$

or $2x_0 I - x_0^2 - a$. The point of this straight line belonging to the horizontal line $\underline{0}$ is $(x_1, 0)$, where

$$x_1 = \frac{1}{2}\left\{x_0 + \frac{a}{x_0}\right\}.$$

It is now geometrically evident why this method of successive approximations serves to define a number whose square is a. Moreover, the geometrical interpretation points the way to the solution of many such problems.

Exercise. If each of a and x_0 is a positive number and, for each positive integer n

$$x_n = \frac{1}{3}\left\{2x_{n-1} + \frac{a}{x_{n-1}^2}\right\},$$

then the sequence x converges to the cube root of a.

Example. If $a = 2$ and $x_0 = 1.26$, then $x_1 = 1.25992$ and $x_1^3 = 1.99996$.

Lemmas Concerning Sequences

1. *Suppose k is a number and for each positive integer n, x_n is a number such that $x_n \leq x_{n+1} \leq k$. There exists a number r with the property that if c is a positive number then there is a positive integer N such that if n > N then x_n differs from r by less than c. That is, the sequence x converges to the sequential limit r.*

2. *Suppose k is a number and for each positive integer n, y_n is a number such that if $y_0 = 0$ then*

$$\sum_{p=1}^{n} |y_p - y_{p-1}| < k.$$

Then there exists a number r such that the sequence y converges to r.

3. *If for each positive integer n, y_n is a number such that $|y_{n+2} - y_{n+1}| \leq (1/2)|y_{n+1} - y_n|$, then there exists a number r such that the sequence y converges to r.*

4. *Suppose that for each positive integer n, y_n is a number such that if c is a positive number then there exists a number N such that y_m differs from y_n by less than c if $m > N$ and $n > N$. Then there exists a number r such that y converges to r.*

5. *Suppose that for each positive integer n, f_n is a simple graph whose X-projection is the set of all numbers, continuous at each of its points and, if c is a positive number and $[a, b]$ is an interval, there exists a positive integer N such that if m and n are integers greater than N and x is a number in $[a, b]$, then*

$$|f_m(x) - f_n(x)| < c.$$

Then there exists a simple graph f whose X-projection is the set of all numbers, continuous at each of its points, such that if c is a positive number and $[a, b]$ an interval then there exists a positive integer N such that if n is an integer greater than N and x is a number in $[a, b]$ then

$$|f(x) - f_n(x)| < c.$$

See Figure 12.2. (Note: c's are lengths of vertical segments.)

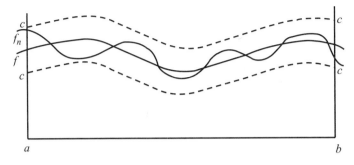

Figure 12.2

Simple Graphs Determined by Successive Approximations

Problem. Suppose (a, b) is a point. We want to establish by successive approximations the existence of only one simple graph f whose X-projection is the set of all numbers such that f contains the point (a, b) and $f' = f$.

Theorems.

1. *If (a, b) is a point and f is a simple graph whose X-projection is the set of all numbers, then these two statements are equivalent*

    (i) $f(a) = b$ and $f' = f$,

    and

    (ii) if x is a number, $f(x) = b + \int_a^x f$.

    This theorem shows that our problem is equivalent to the problem of showing that there exists only one simple graph f for which (ii) is true.

2. *Suppose f_0 is a simple graph whose X-projection is the set of all numbers continuous at each of its points, and for each positive integer n and number x,*

$$f_n(x) = b + \int_a^x f_{n-1}.$$

   Then f_n is continuous at each of its points and $f_{n+1}(x) - f_n(x) = \int_a^x (f_n - f_{n-1}).$

3. *If $[A, B]$ is an interval containing the number a then there exists a number M such that if x is in $[A, B]$ then*

$$|f_1(x) - f_0(x)| \leq M$$

and for each positive integer n,

$$|f_{n+1}(x) - f_n(x)| \leq M \cdot \frac{|x - a|^n}{1 \cdot 2 \cdots n}.$$

4. *If c is a positive number, there exists a positive integer N such that if x is in $[A, B]$, then $f_m(x)$ differs from $f_n(x)$ by less than c if $m > N$ and $n > N$.*

5. *There exists a simple graph f whose X-projection is the set of all numbers such that if $[A, B]$ is an interval and c is a positive number then there exists a positive integer N such that if x is in $[A, B]$ and n is an integer greater than N, then $|f(x) - f_n(x)| < c$.*

6. *f is continuous at each of its points.*

7. *If for each positive integer n, ε_n is the simple graph defined by $f_n - f = \varepsilon_n$, so that $f_n = f + \varepsilon_n$ and if ε_n is continuous at each of its points, then if x is a number we have*

$$f(x) - \left\{ b + \int_a^x f \right\} = \int_a^x \varepsilon_{n-1} - \varepsilon_n.$$

The assumption that there is a number x such that $|f(x) - \{b + \int_a^x f\}|$ is a positive number δ leads to a contradiction so that for every number x, $f(x) = b + \int_a^x f$.

8. *If F is a simple graph whose X-projection is the set of all numbers such that if x is a number then $F(x) = b + \int_a^x F$, then $F(x) - f(x) = \int_a^x (F - f)$ and it follows that $F = f$.*

The truth of these statements establishes the following theorem.

Theorem. *If (a, b) is a point then there exists only one simple graph f containing the point (a, b) such that for every number x, $f'(x) = f(x)$.*

Corollary. *Suppose E denotes the simple graph f whose X-projection is the set of all numbers such that $f(0) = 1$ and $f' = f$. Then the following statements are true.*

(i) *If x is a number, $E(x) > 0$.*

(ii) *If x and y are numbers and $x < y$, then $E(x) < E(y)$.*

(iii) *If* f *is* the *simple graph with* X-*projection the set of all numbers containing the point* (a, b) *such that* $f' = f$, *then* $f = bE[I - a]$.

(iv) *If each of* c *and* x *is a number,* $E(x + c) = E(x) \cdot E(c)$.

Theorems.

1. *If* (a, b) *is a point and* g *is a simple graph whose* X-*projection is the set of all numbers, continuous at each of its points, then there exists a simple graph* u *such that the only simple graph* f *containing* (a, b) *for which* $f' - f = g$ *is* $uE[I - a]$.

2. *If* (a, b) *is a point and each of* q *and* g *is a simple graph whose* X-*projection is the set of all numbers, continuous at each of its points, then there exists a simple graph* h *and a simple graph* u *such that the only simple graph* f *containing* (a, b) *for which* $f' - qf = g$ *is* $uE[h - h(a)]$.

Exercise. For each nonnegative integer n, suppose E_n denotes the simple graph

$$E - \left\{ 1 + \frac{I}{1} + \frac{I^2}{1 \cdot 2} + \cdots + \frac{I^n}{1 \cdot 2 \cdot \cdots \cdot n} \right\}.$$

Use the preceding result to determine E_n and show that if x is a positive number then

$$0 < E_n(x) < E(x) \cdot \frac{x^{n+1}}{1 \cdot 2 \cdot \cdots \cdot (n+1)}.$$

The Equation $f' - f = -H$

Problem. Show that there is only one simple graph f whose X-projection is the set of all positive numbers such that $f' - f = -H$ and that, if $\underline{\alpha}$ and β are horizontal lines with $\underline{0}$ between them, there exists a vertical line $h|$ such that every point of f to the right of $h|$ is between $\underline{\alpha}$ and $\underline{\beta}$. Show, moreover, that if x is a positive number and n a positive integer, the difference

$$f(x) - \left\{ \frac{1}{x} - 1 \cdot \left(\frac{1}{x}\right)^2 + 1 \cdot 2 \cdot \left(\frac{1}{x}\right)^3 - \cdots + (-1)^n \cdot 1 \cdot 2 \cdot \cdots \cdot n \cdot \left(\frac{1}{x}\right)^{n+1} \right\}$$

is positive or negative according as n is odd or even, it is numerically less than

$$1 \cdot 2 \cdot \cdots \cdot (n+1) \cdot \left(\frac{1}{x}\right)^{n+2},$$

and

$$-\left(\frac{1}{x}\right)^2 < f(x) - \frac{1}{x} < 0.$$

Note.

$$\left| f(x) - \frac{1}{x} \right| < .005, \qquad\qquad \text{if } x > 14.2;$$

$$\left| f(x) - \left\{ \frac{1}{x} - \left(\frac{1}{x} \right)^2 \right\} \right| < .005, \qquad\qquad \text{if } x > 7.5;$$

$$\left| f(x) - \left\{ \frac{1}{x} - \left(\frac{1}{x} \right)^2 + 2 \cdot \left(\frac{1}{x} \right)^3 \right\} \right| < .005, \quad \text{if } x \geq 6.$$

Exercise. Find a way to compute approximations to $f(x)$ correct to two decimal places in the range $0 < x < 6$.

Ordered Pairs of Simple Graphs Determined by Successive Approximations

Problem. Suppose each of a, b, and c is a number and each of q and r is a simple graph with X-projection the set of all numbers, continuous at each of its points. Show that there exists *only one* ordered pair (f, g), each member of which is a simple graph with X-projection the set of all numbers, such that

$$f(a) = b, \quad f' = qg,$$
$$g(a) = c, \quad g' = rf.$$

The pattern is similar to that in the preceding problem. The problem is equivalent to the problem of determining an ordered pair (f, g), such that if x is a number then

$$f(x) = b + \int_a^x qg,$$

$$g(x) = c + \int_a^x rf.$$

We leave the details to the reader.

Theorem. *Suppose each of a, b, and c is a number and each of q and r is a simple graph with X-projection the set of all numbers, continuous at each of its points. If t is a number, denote by (A_t, C_t) the ordered pair (f, g), such that*

$$f(t) = 1, \quad g(t) = 0, \quad f' = qg, \quad g' = rf,$$

and denote by (B_t, D_t) *the ordered pair* (f, g) *such that*

$$f(t) = 0, \quad g(t) = 1, \quad f' = qg, \quad g' = rf,$$

so that we have

$$A_t(t) = 1, \quad B_t(t) = 0, \quad A_t' = qC_t, \quad B_t' = qD_t,$$
$$C_t(t) = 0, \quad D_t(t) = 1, \quad C_t' = rA_r, \quad D_t' = rB_t.$$

The following statements may be established as consequences of the above definitions and the fact that there is *only one* ordered pair (f, g) containing the points (a, b) and (a, c), respectively, such that $f' = qg$ and $g' = rf$.

i. If t is a number, $A_t D_t - B_t C_t = \underline{1}$.

ii. The ordered pair (f, g) containing (a, b) and (a, c), respectively, such that $f' = qg$ and $g' = rf$ is $(bA_a + cB_a, bC_a + cD_a)$.

iii. If (x, y, z) is an ordered number triple,

$$A_z(x) = A_y(x)A_z(y) + B_y(x)C_z(y),$$
$$C_z(x) = C_y(x)A_z(y) + D_y(x)C_z(y),$$

and

$$B_z(x) = A_y(x)B_z(y) + B_y(x)D_z(y),$$
$$D_z(x) = C_y(x)B_z(y) + D_y(x)D_z(y).$$

iv. If (x, t) is an ordered number pair, $A_t(x) = D_x(t)$, $B_t(x) = -B_x(t)$, and $C_t(x) = -C_x(t)$.

v. If each of a, b, and c is a number and each of u and v is a simple graph whose X-projection is the set of all numbers, continuous at each of its points, then the only ordered pair (f, g) such that $f(a) = b$, $g(a) = c$, $f' - qg = u$, and $g' - rf = v$ is given, for each number x, by

$$f(x) = bA_a(x) + cB_a(x) + \int_a^x \{uD_x - vB_x\},$$

$$g(x) = bC_a(x) + cD_a(x) + \int_a^x \{-uC_x - vA_x\}.$$

Particular Examples

The rest of this chapter contains outlines that suggest investigations of particular cases of the ideas of the last section.

(A) The Simple Graphs S and C

Theorem. *Suppose* (A_t, B_t, C_t, D_t) *is the ordered quadruple of the preceding section with* $q = -1$ *and* $r = 1$, *so that*

$$A_t(t) = 1, \quad B_t(t) = 0, \quad A_t' = -C_t, \quad B_t' = -D_t,$$

$$C_t(t) = 0, \quad D_t(t) = 1, \quad C_t' = A_t, \quad D_t' = B_t.$$

Then

i. If t is a number then $A_t = D_t$ and $B_t = -C_t$.

ii. If (x, y, z) is an ordered number triple then

$$A_z(x) = A_y(x)A_z(y) - B_y(x)B_z(y),$$

$$B_z(x) = A_y(x)B_z(y) + B_y(x)A_z(y).$$

iii. If h is a number, then for every number x and number z,

$$A_{z+h}(x + h) = A_z(x),$$

$$B_{z+h}(x + h) = B_z(x).$$

That is, if (x, z) is an ordered number pair, (x', z') is an ordered number pair and $x' - x = z' - z$, then

$$A_{z'}(x') = A_z(x) \quad \text{and} \quad B_{z'}(x') = B_z(x).$$

This means that there is a simple graph C and a simple graph S such that if (x, t) is an ordered number pair then

$$A_t(x) = C(x - t) \quad \text{and} \quad B_t(x) = S(x - t).$$

iv. $C^2 + S^2 = 1$.

v. If u is a number and v is a number,

$$C(u + v) = C(u)C(v) - S(u)S(v),$$

$$S(u + v) = S(u)C(v) + C(u)S(v).$$

vi. $C' = -S$ and $S' = C$.

vii. There exists a positive number r such that $C(r) = 0$. There is a least such positive number r, which we denote by $\pi/2$.

viii. If x is a number then $C(x + 2\pi) = C(x)$ and $S(x + 2\pi) = S(x)$.

ix. Suppose that for each positive integer n,

$$S = I - \frac{I^3}{1 \cdot 2 \cdot 3} + \cdots + (-1)^{n-1} \frac{I^{2n-1}}{1 \cdot 2 \cdots (2n-1)} + S_n$$

and

$$C = 1 - \frac{I^2}{1 \cdot 2} + \frac{I^4}{1 \cdot 2 \cdot 3 \cdot 4} - \cdots + (-1)^n \frac{I^{2n}}{1 \cdot 2 \cdots (2n)} + C_n.$$

Show that

$$C_n(0) = 0, \quad C_n' + S_n = \underline{0},$$

$$S_n(0) = 0, \quad S_n' - C_n = (-1)^n \frac{I^{2n}}{1 \cdot 2 \cdots (2n)}$$

and hence express S_n and C_n as integrals. (See (v) of the preceding section.)

(B) The Equation $f' = I^2 + f^2$

Problem. If (a, b) is a point, determine a simple graph f containing (a, b) whose X-projection is some segment containing a such that the slope of f at each of its points is the square of the radius of the circle with center $(0, 0)$ that contains that point.

In Figure 12.3, (a, b) is $(0, 0)$ and the simple graph f that is sketched in has slope at each of its points approximately the square of the radius of the circle with center $(0, 0)$ containing that point. That is, at A the slope of f is $1/4$, at B 1, at C 9/4, $\ldots$.

The fact that $T' = 1 + T^2$ and $T = S/C$ suggests, by analogy, that we try to express the simple graph f containing $(0, 0)$ such that if (x, y) is a point of f then $f'(x) = x^2 + y^2$, as the quotient u/v of two simple graphs. We would then have

$$\frac{vu' - uv'}{v^2} = I^2 + \left(\frac{u}{v}\right)^2 = \frac{I^2 v^2 + u^2}{v^2}$$

and, to ensure that $f(0) = 0$,

$$u(0) = 0, \quad v(0) = 1.$$

The conditions are met if

(i)
$$u(0) = 0, \quad u' = I^2 v,$$
$$v(0) = 1, \quad v' = -u.$$

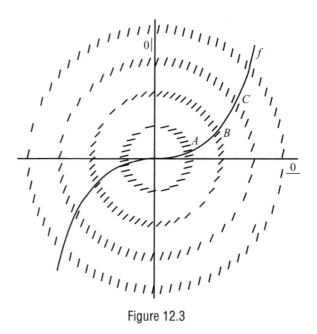

Figure 12.3

This is the case $q = I^2$ and $r = -\underline{1}$ of our general system. In the notation previously introduced, a simple graph f with the desired properties is the quotient B_0/D_0 for the system (i).

Theorem. *Suppose (a, b) is a point and each of q and r is a simple graph with X-projection the set of all numbers, continuous at each of its points. There exists a segment s containing a such that the only simple graph f with X-projection s containing the point (a, b) for which*

$$f' = q - rf^2$$

is

$$\frac{b \cdot A_a + B_a}{b \cdot C_a + D_a},$$

where (A_a, B_a, C_a, D_a) is the ordered quadruple previously introduced (for q and r used here).

Theorem. *Denote by $(\alpha, \beta, \gamma, \delta)$ the ordered quadruple (A_t, B_t, C_t, D_t) for the system (i) in case $t = 0$. Then* *i. For each number x,*

$$\alpha(x) = 1 + \int_0^x I^2\gamma, \quad \beta(x) = \int_0^x I^2\delta,$$

$$\gamma(x) = -\int_0^x \alpha, \quad \delta(x) = 1 - \int_0^x \beta.$$

ii. If t is a number,

$$A_t = \alpha \cdot \delta(t) - \beta \cdot \gamma(t), \quad B_t = \gamma \cdot \delta(t) - \delta \cdot \gamma(t),$$
$$C_t = \beta \cdot \alpha(t) - \alpha \cdot \beta(t), \quad D_t = \delta \cdot \alpha(t) - \gamma \cdot \beta(t).$$

iii. If n is a positive integer,

$$\alpha = 1 - \frac{I^4}{1 \cdot 4} + \frac{I^8}{(1 \cdot 4)(5 \cdot 8)} - \frac{I^{12}}{(1 \cdot 4)(5 \cdot 8)(9 \cdot 12)} + \cdots$$

$$+ (-1)^n \frac{I^{4n}}{(1 \cdot 4)(5 \cdot 8) \cdots ([4n - 3][4n])} + \alpha_n,$$

where, if x is a number distinct from 0 and n a positive integer such that

$$\frac{x^4}{(4n - 3)(4n)} < 1,$$

$$|\alpha(x) - \alpha_n(x)| < \frac{x^{4n+4}}{(1 \cdot 4)(5 \cdot 8) \cdots ([4n + 1][4n + 4])}.$$

With an analogous estimate for the error in each case, we have

$$\beta = \frac{I^3}{3} - \frac{I^7}{3(4 \cdot 7)} + \frac{I^{11}}{3(4 \cdot 7)(8 \cdot 11)} - \frac{I^{15}}{3(4 \cdot 7)(8 \cdot 11)(12 \cdot 15)} + \cdots$$

$$+ (-1)^n \frac{I^{4n+3}}{3(4 \cdot 7)(8 \cdot 11) \cdots (4n[4n + 3])} + \beta_n,$$

$$\gamma = -I + \frac{I^5}{4 \cdot 5} - \frac{I^9}{(4 \cdot 5)(8 \cdot 9)} + \frac{I^{13}}{(4 \cdot 5)(8 \cdot 9)(12 \cdot 13)} - \cdots$$

$$+ (-1)^{n-1} \frac{I^{4n+1}}{(4 \cdot 5)(8 \cdot 9) \cdots (4n[4n + 1])} + \gamma_n,$$

$$\delta = 1 - \frac{I^4}{3 \cdot 4} + \frac{I^8}{(3 \cdot 4)(7 \cdot 8)} - \frac{I^{12}}{(3 \cdot 4)(7 \cdot 8)(11 \cdot 12)} + \cdots$$

$$+ (-1)^n \frac{I^{4n}}{(3 \cdot 4)(7 \cdot 8) \cdots ([4n - 1][4n])} + \delta_n.$$

Exercise. The simple graph f in Figure 12.3 such that $f(0) = 0$ and $f' = I^2 + f^2$ is β/δ (restricted to a segment about 0).

   i. Show that $f(1) = .35023$ to five decimal places.

   ii. Show that $\delta(2) = .004$. The least positive number r such that $\delta(r) = 0$ is approximately 2. A better approximation to r is 2.00314. The X-projection of the simple graph f is the set of all numbers between $-r$ and r.

   iii. Indicate a way to compute points of the simple graph f containing the point (a, b) such that $f' = I^2 + f^2$.

   iv. Show that $\delta'' + I^2\delta = \underline{0}$, $\gamma'' + I^2\gamma = \underline{0}$, $\gamma\delta' - \delta\gamma' = \underline{1}$.

   v. Suppose each of a, b, and c is a number and each of p, q, r, and s a simple graph with X-projection the set of all numbers, continuous at each of its points. Define h and k by

$$h(x) = -\int_a^x p \quad \text{and} \quad k(x) = -\int_a^x s,$$

F and G by

$$F = E[h] \cdot f \quad \text{and} \quad G = E[k] \cdot g,$$

and q_1 and r_1 by

$$q_1 = q \cdot E[h - k] \quad \text{and} \quad r_1 = r \cdot E[k - h].$$

The following two statements are equivalent.

   i. $f(a) = b$, $\quad g(a) = c$, $\quad f' = pf + qg$, $\quad g' = rf + sg$,

   and

   ii. $F(a) = b$, $\quad G(a) = c$, $\quad F' = q_1G$, $\quad$ and $\quad G' = r_1F$.

Thus, the solution of the system (i) is equivalent to the solution of the system (ii).

(C) The Equation $uf'' + vf' + wf = g$

Problem. Suppose each of u, v, w, and g is a simple graph whose X-projection is the set of all numbers, continuous at each of its points and if x is a number then $u(x) > 0$. If each of a, b, and c is a number, determine a simple graph f such that $f(a) = b$, $f'(a) = c$, and $uf'' + vf' + wf = g$.

Theorems. *The following statements are true under the conditions just stated.*

1. *There exists a simple graph m, continuous at each of its points, such that if x is a number, then $m(x) > 0$; a simple graph p, continuous at each of its points, such that, if x is a number, then $p(x) > 0$; and a simple graph q, continuous at $(x, q(x))$ for every number x, such that for appropriate f,*

$$m\{uf'' + vf' + wf\} = \{pf'\}' - qf,$$

so that the above problem may be replaced by the equation

$$\{pf'\}' - qf = mg.$$

2. *The last equation with the conditions $f(a) = b$, $f'(a) = c$, is a system equivalent to*

$$f(a) = b, \qquad f' = \frac{1}{p} \cdot g,$$

$$g(a) = cp(a), \qquad g' = q \cdot f + mg,$$

which is a particular case of the system considered in the preceding sections.

3. *Suppose each of f and g is a simple graph, $f'' - qf = \underline{0}$, and, $g'' - qg = \underline{0}$. Then*

   i. *There exists a number c such that $fg' - gf' = \underline{c}$.*

   ii. *If $c \neq 0$, then if x is a number,*

$$|f(x)| + |g(x)| > 0,$$

$$|f'(x)| + |g'(x)| > 0,$$

$$|f(x)| + |f'(x)| > 0.$$

   iii. *If $c \neq 0$ and M is the number set to which x belongs only if $f(x) = 0$, then no interval includes an infinite subset of M.*

   iv. *If $c \neq 0$ and (h, k) is an ordered number pair such that $hf + kg = 0$, then $h = 0$ and $k = 0$.*

   v. *If $c = 0$, then there exists a number k such that $f = kg$, or a number k such that $g = kf$.*

   vi. *If $c \neq 0$ and r and s are numbers such that $f(r) = f(s) = 0$, and if x is between r and s, $f(x) \neq 0$, then there exists only one number t between r and s such that $g(t) = 0$.*

vii. *If Q is a continuous simple graph such that $Q(x) \geq q(x)$ for every number x and each segment contains a number x such that $Q(x) > q(x)$, if $f'' + qf = \underline{0}$, $F'' + QF = \underline{0}$, and r and s are numbers such that $f(r) = 0$ and $f(s) = 0$, then there is a number t between r and s such that $F(t) = 0$.*

13

Linear Spaces of Simple Graphs

In this chapter we outline some selected theorems about the space of points, i.e., the plane of ordered number pairs, and then try to lead the reader to extend them to a space in which the word "point" is interpreted to mean "continuous simple graph with X-projection the interval $[a, b]$," i.e., a space in which "point" means a certain kind of set of ordinary points.

Linear Transformations in the Plane

To make the extension easier to accomplish, we make some changes in notation.

If f is a point, i.e., an ordered number pair, we denote the point by $(f(1), f(2))$. (A *point* f is determined by $f(x)$, $x = 1, 2$, while a simple graph with X-projection $[a, b]$ is determined by $f(x)$, $a \leq x \leq b$.)

Definition. The statement that the transformation W from the set of all points to a point set *is linear* means that if f is a point, g is a point, and k is a number, then

$$W(f + g) = Wf + Wg \quad \text{and} \quad W(k \cdot f) = k \cdot Wf.$$

(See the definitions of *sum, product by a number, inner product*, and *absolute value* for points on page 72.)

135

Theorems. *Suppose W is a transformation from the set of all points to a point set.*

1. *The following two statements are equivalent.*

    (i) W is linear.

    (ii) There exists a transformation K from the set of four points $\{(1, 1), (1, 2), (2, 1), (2, 2)\}$ to a number set such that, if h is a point and $f = Wh$, then

    $$f(x) = \sum_{y=1}^{2} K(x, y)h(y), \quad x = 1, 2.$$

2. *If W is linear, then either there exists a point f distinct from $(0, 0)$ such that $Wf = (0, 0)$ or, if g is a point, there exists only one point f such that $Wf = g$.*

3. *If W is linear and reversible and if $K(1, 2) = K(2, 1)$, i.e., K is symmetric, then there exists an order number pair (λ_1, λ_2) and an ordered point pair (ϕ_1, ϕ_2) such that*

    (i) $|\phi_1| = 1, |\phi_2| = 1$, and $((\phi_1, \phi_2)) = 0$,

    and

    (ii) $\phi_1 = \lambda_1 \cdot W\phi_1$ and $\phi_2 = \lambda_2 \cdot W\phi_2$.

    Also,

    $$K(x, y) = \sum_{p=1}^{2} \frac{\phi_p(x)\phi_p(y)}{\lambda_p}, \quad x = 1, 2; \quad y = 1, 2;$$

    and if ϕ is a point,

    $$\phi = \sum_{p=1}^{2} ((\phi, \phi_p)) \cdot \phi_p.$$

Example. If $K(1, 1) = 5$, $K(1, 2) = K(2, 1) = 2$, and $K(2, 2) = 2$, then

$$\lambda_1 = 1, \ \lambda_2 = \frac{1}{6}, \ \phi_1 = \left(-\frac{1}{\sqrt{5}}, \frac{2}{\sqrt{5}}\right), \text{ and } \phi_2 = \left(\frac{2}{\sqrt{5}}, \frac{1}{\sqrt{5}}\right).$$

On the other hand, if $K(1, 1) = 1$, $K(1, 2) = K(2, 1) = 0$, and $K(2, 2) = 1$, then $\lambda_1 = \lambda_2 = 1$, $\phi_1 = (\frac{3}{5}, \frac{4}{5})$, and $\phi_2 = (\frac{4}{5}, -\frac{3}{5})$.

Axioms for Linear Space

The statement that S is a *linear space* means that S consists of a point set, where the word *point* is undefined, in which there is an addition and a multiplication by numbers satisfying the following axioms.

Axiom I.

i. If f is a point and g is a point, then there is only one point $f + g$ (called the *sum* of f and g).

ii. If each of f, g, and h is a point, then

$$f + (g + h) = (f + g) + h.$$

iii. If f is a point and g is a point, then

$$f + g = g + f.$$

iv. There is a point N such that if f is a point, then

$$N + f = f.$$

Axiom II.

i. If f is a point and k is a number, then there is only one point $k \cdot f$ (called the *product k dot f*).

ii. If each of f and g is a point and k is a number, then

$$k \cdot (f + g) = k \cdot f + k \cdot g.$$

iii. If each of k_1 and k_2 is a number and f is a point, then

$$k_1 \cdot (k_2 \cdot f) = (k_1 k_2) \cdot f.$$

iv. If each of k_1 and k_2 is a number and f is a point, then

$$(k_1 + k_2) \cdot f = k_1 \cdot f + k_2 \cdot f.$$

Axiom III. If k is a number and f is a point, then the following two statements are equivalent.

i. $k \cdot f = N$.

ii. $k = 0$ or $f = N$.

Theorems.

1. If N' is a point such that for every point f,

$$N' + f = f,$$

then $N' = N$.

The point N is called the *zero point* of S.

2. If f is a point, then

$$1 \cdot f = f.$$

The point $-1 \cdot f$ is denoted by $-f$ and called the *negative* of the point f. We have

$$f + (-f) = 1 \cdot f + (-1) \cdot f = [1 + (-1)] \cdot f = 0 \cdot f = N.$$

If each of f and g is a point, the point $f + (-g)$ is denoted by $f - g$ and called the *difference f minus g*. In particular, $f - f = N$.

The statement that S is a *normed* linear space means that S is a linear space such that for each point f there is a number $\|f\|$ (called the *norm* of f) such that

    i. if f is a point, then $\|f\| > 0$ unless $f = N$,

    ii. if f is a point and k is a number, then $\|k \cdot f\| = |k| \, \|f\|$, and

    iii. if each of f and g is a point, then

$$\|f + g\| \le \|f\| + \|g\|.$$

If each of f and g is a point, $\|f - g\|$ is the *distance* from f to g. Distance has the properties

    i. if each of f and g is a point, then $\|f - g\| = \|g - f\|$ and $\|f - g\| > 0$ unless $f = g$,

    ii. if each of f and g is a point and k is a number, then

$$\|k \cdot f - k \cdot g\| = |k| \, \|f - g\|,$$

    and

    iii. if each of f, g, and h is a point, then

$$\|f - h\| \le \|f - g\| + \|g - h\|.$$

The statement that S is an *inner product* space means that S is a linear space such that for each point f and point g there is a number $((f, g))$ (called the inner product of f and g) with the properties

    i. if each of f and g is a point, then $((f, g))$ is a number and $((f, f)) > 0$ unless $f = N$,

    ii. if each of f and g is a point, then

$$((f, g)) = ((g, f)),$$

    iii. if each of f and g is a point and k is a number, then

$$((k \cdot f, g)) = k((f, g)),$$

    and

    iv. if each of f, g, and h is a point, then

$$((f + g, h)) = ((f, h)) + ((g, h)).$$

Theorem. *If each of f and g is a point, then*

$$((f, g))^2 \leq ((f, f))((g, g)).$$

The inner product space is a normed linear space with the norm of the point f given by

$$\|f\| = ((f, f))^{1/2}.$$

Examples.

1. The space S in which *point* means *number*, with ordinary addition and multiplication, is an inner product space with the inner product the ordinary product.

2. The plane in which *point* means ordered number pair, with ordinary addition and multiplication by numbers, is an inner product space if inner product is defined as on page 72.

3. The space in which *point* means simple graph with X-projection the interval $[a, b]$, with ordinary addition and multiplication by numbers, is a linear space.

The Inner Product Space $M[a, b]$

In the rest of this chapter the simple graphs considered have X-projection an interval or, in one case, a segment. We shall agree that a familiar symbol for a particular simple graph, such as $I, L, E, S, K, \underline{k}$, shall represent the subset of the simple graph formerly represented by that symbol whose X-projection is the interval or segment used. For instance, if the interval $[0, 1]$ is the X-projection of all simple graphs used in a certain discussion then, in that discussion, $\underline{0}$ denotes the simple graph with X-projection $[0, 1]$ such that if x is in $[0, 1]$ then $f(x) = 0$.

Definition. $M[a, b]$ denotes the inner product space in which *point* means continuous simple graph with X-projection the interval $[a, b]$ with inner product defined by

$$((f, g)) = \int_a^b fg$$

for every f and g in $M[a, b]$. The *norm* is then $\{\int_a^b f^2\}^{1/2} = \|f\|$ and the *distance* from f to g is $\|f - g\|$.

Some Linear Transformations in $M[a, b]$

Suppose each of p and q is a continuous simple graph with X-projection $[a, b]$ and if x is in $[a, b]$ then

$$p(x) > 0.$$

Denote by $M_0[a, b]$ the inner product space in which *point* means an f belonging to $M[a, b]$ such that

$$f(a) = 0, \quad f(b) = 0$$

and

$$qf - \{pf'\}'$$

belongs to $M[a, b]$, with the inner product defined in $M[a, b]$. That is, $M_0[a, b]$ is a *subspace* of $M[a, b]$.

The transformation V from $M_0[a, b]$ to a point set of $M[a, b]$ defined for each f in $M_0[a, b]$ by

$$Vf = qf - \{pf'\}'$$

is *linear*, so if f is in $M_0[a, b]$, g is in $M_0[a, b]$, and k is a number, then

$$V(f + g) = Vf + Vg, \quad \text{and} \quad V(k \cdot f) = k \cdot Vf.$$

By restricting all simple graphs considered to have X-projection $[a, b]$, we have from the work of the preceding chapter, for each t in $[a, b]$, the ordered quadruple (A_t, B_t, C_t, D_t) each member of which is a simple graph with X-projection $[a, b]$, defined by

$$A_t(t) = 1, \quad B_t(t) = 0, \quad A_t' = \frac{1}{p} C_t, \quad B_t' = \frac{1}{p} D_t,$$

$$C_t(t) = 0, \quad D_t(t) = 1, \quad C_t' = q A_t, \quad D_t' = q B_t.$$

We state the formulas for convenience in reference.

$$A_t D_t - B_t C_t = \underline{1},$$

$$A_t(x) = D_x(t), \quad B_t(x) = -B_x(t), \quad \text{and} \quad C_t(x) = -C_x(t),$$

$$A_y(x) = A_t(x)A_y(t) + B_t(x)C_y(t),$$
$$C_y(x) = C_t(x)A_y(t) + D_t(x)C_y(t),$$
$$B_y(x) = A_t(x)B_y(t) + B_t(x)D_y(t),$$
$$D_y(x) = C_t(x)B_y(t) + D_t(x)D_y(t).$$

Also,

$$V A_t = \underline{0}, \quad A_t(t) = 1, \quad A_t'(t) = 0,$$
$$V B_t = \underline{0}, \quad B_t(t) = 0, \quad p(t)B_t'(t) = 1.$$

Theorem. *There are two alternatives: either*

i. *there exists a point f in $M_0[a, b]$, distinct from $\underline{0}$, such that $Vf = \underline{0}$, or*

ii. *if h is in $M[a, b]$, there exists only one f in $M_0[a, b]$ such that $Vf = h$.*

The first alternative occurs when $B_a(b) = 0$ and the second when $B_a(b) \neq 0$. In the second V is a reversible transformation from $M_0[a, b]$ to $M[a, b]$. If W is the inverse of the transformation V and h is in $M[a, b]$, then the only f in $M_0[a, b]$ such that $Vf = h$, i.e., Wh, is given by

$$f(x) = \int_a^b K[x, I]h, \quad a \leq x \leq b,$$

where K is the transformation from $[ab; ab]$ to a number set defined by

$$K(x, y) = \begin{cases} \dfrac{B_a(x)B_b(y)}{B_b(a)}, & a \le x \le y \le b, \\[2ex] \dfrac{B_a(y)B_b(x)}{B_b(a)}, & a \le y \le x \le b. \end{cases}$$

K is called the *kernel* of the transformation W. Note that K is *symmetric*: if (x, y) is in $[ab; ab]$, $K(x, y) = K(y, x)$.

Exercise. If $[a, b] = [0, 1]$, $Vf = -f''$, show that the second alternative holds and that

$$K(x, y) = \begin{cases} x(1 - y), & 0 \le x \le y \le 1, \\ y(1 - x), & 0 \le y \le x \le 1. \end{cases}$$

Recipe for the Kernel

We state a formula that furnishes a key to other problems of the kind just considered.

Theorem. *Suppose f and g are simple graphs such that Vf and Vg are integrable on the subinterval $[s, t]$ of $[a, b]$. Then*

$$\int_s^t \{gVf - fVg\} = \{p(fg' - gf')\}\big|_s^t.$$

The simple graphs f that are the points of the space $M_0[a, b]$ are required to satisfy the *endpoint* conditions $f(a) = 0$, $f(b) = 0$. Denote by c a pair of endpoint conditions such that if f and g are simple graphs with X-projection $[a, b]$ satisfying the endpoint conditions c, then

$$\{p(fg' - gf')\}\big|_a^b = 0.$$

Denote by $M_c[a, b]$ the subspace of $M[a, b]$ defined in the same way $M_0[a, b]$ was defined except with conditions c instead of with the conditions $f(a) = 0$, $f(b) = 0$.

Examples of admissible endpoint conditions are

$$f'(a) = 0, \quad f'(b) = 0,$$

and

$$f(a) = -f(b), \quad p(a)f'(a) = -p(b)f'(b).$$

Definition. The statement that K is a *kernel* of the transformation V under the endpoint conditions c means that K is a transformation from $[ab; ab]$ to a number set such that if y is a number between a and b, u is the subset of $K[I, y]$ with X-projection $[a, y]$, and v is the subset of $K[I, y]$ with X-projection $[y, b]$, then

    i. $Vu = \underline{0}$ and $Vv = \underline{0}$,

    ii. $u'(y) - v'(y) = \dfrac{1}{p(y)}$,

    iii. $K[I, y]$ is continuous, and

    iv. $K[I, y]$ satisfies the endpoint conditions c.

Note. This recipe for K may be stated in what is perhaps a self-explanatory notation as follows:

    i. $VK[I, y] = \underline{0}$ on $[a, y]$ and on $[y, b]$,

    ii. $K_1'(y-, y) - K_1'(y+, y) = \dfrac{1}{p(y)}$,

    iii. $K(y-, y) = K(y+, y)$, and

    iv. $K[I, y]$ satisfies conditions c.

Example. Let us construct K according to the recipe in case $[a, b] = [0, 1]$, $Vf = -f''$, and c is $f(0) = 0$, $f(1) = 0$.

If y is a number between 0 and 1, then by (i),

$$K(x, y) = \begin{cases} c_1 x + c_2, & 0 \le x \le y, \\ c_3 x + c_4, & y \le x \le 1, \end{cases}$$

where each of c_1, c_2, c_3, and c_4 is a number. By (ii),

$$c_1 - c_3 = 1,$$

by (iii)

$$c_1 y + c_2 = c_3 y + c_4,$$

and by (iv)

$$c_2 = 0 \quad \text{and} \quad c_3 + c_4 = 0.$$

These four equations determine c_1, c_2, c_3, c_4:

$$c_1 = 1 - y, \quad c_2 = 0, \quad c_3 = -y, \quad c_4 = y,$$

and so

$$K(x, y) = \begin{cases} x(1 - y), & 0 \le x \le y, \\ y(1 - x), & y \le x \le 1. \end{cases}$$

Theorem. *Suppose each of K and k is a kernel of the transformation V under the endpoint conditions c. Then the following statements are true.*

  i. *If (x, y) is a point interior to $[ab; ab]$, then*

$$K(x, y) = k(y, x),$$

K is symmetric, i.e., $K(x, y) = K(y, x)$, and $k = K$.

  ii. *If h is in $M[a, b]$ and if there is an f in $M_c[a, b]$ such that $Vf = h$, then*

$$f(x) = \int_a^b K[x, I]h, \quad a \le x \le b.$$

Exercise. Suppose $Vf = -f''$. Find the kernel of V under the given endpoint conditions:

  i. $[a, b] = [0, 1]$, $f(0) = 0$, $f'(1) = 0$.

  ii. $[a, b] = [-1, 1]$, $f(-1) = 0$, $f(1) = 0$.

  iii. $[a, b] = [0, 1]$, $f(0) = -f(1)$, $f'(0) = -f'(1)$.

In each case V is a reversible transformation from $M_c[a, b]$ to $M[a, b]$ with inverse transformation defined by

$$f(x) = \int_a^b K[x, I]h, \quad a \le x \le b.$$

Suppose c is a pair of endpoint conditions such that in addition to the conditions already imposed, if f satisfies c, then $pff'\,|_a^b = 0$, and suppose V is a reversible transformation from $M_c[a, b]$ to $M[a, b]$ with kernel K. If h is in $M[a, b]$ and $Vf = h$, then

$$\int_a^b \int_a^b K(x, y)h(x)h(y)dxdy = \int_a^b \{pf'^2 + qf^2\}.$$

This formula shows that if $q(x) \ge 0$ for every x in $[a, b]$, then $\int_a^b \int_a^b K(x, y)h(x)h(y)dxdy \ge 0$ for every h in $M[a, b]$.

Another Kind of Endpoint Problem

Denote by $M(-1, 1)$ the linear space in which *point* means continuous simple graph with X-projection the segment Δ with ends -1 and 1 and by $M_1(-1, 1)$ the subspace of $M(-1, 1)$ in which *point* means an f in $M(-1, 1)$ such that f'' is continuous and has X-projection the segment Δ.

If we restrict all simple graphs considered to have X-projection Δ, the earlier investigations show that for t in Δ there is an ordered quadruple (A_t, B_t, C_t, D_t) defined by

$$A_t(t) = 1, \quad B_t(t) = 0, \quad A'_t = \frac{1}{1 - I^2}C_t, \quad B'_t = \frac{1}{1 - I^2}D_t,$$

$$C_t(t) = 0, \quad D_t(t) = 1, \quad C'_t = \underline{0}, \quad D'_t = \underline{1},$$

namely

$$A_t = \underline{1}, \quad B_t = \frac{1}{2}L\left[\frac{(1 + I)(1 - t)}{(1 + t)(1 - I)}\right], \quad C_t = \underline{0}, \quad \text{and} \quad D_t = \underline{1}.$$

If each of a, b, and c is a number and a is in Δ, there exists for each h in $M(-1, 1)$ only one f in $M_1(-1, 1)$ such that

$$f(a) = b, \quad f'(a) = c, \quad \text{and} \quad -\{(1 - I^2)f'\}' = h.$$

Thus the transformation Z defined for each f in $M_1(-1, 1)$ by

$$Zf = -\{(1 - I^2)f'\}'$$

is a transformation from $M_1(-1, 1)$ to $M(-1, 1)$.

Problem. Find a subspace G of $M(-1, 1)$ and a subspace G_1 of $M_1(-1, 1)$ such that Z is a reversible transformation from G_1 to G.

Note 1. If f is in $M(-1, 1)$ and has derivative f' bounded on the common part of Δ and a segment containing -1 and on the common part of Δ and a segment containing 1, then f is bounded on Δ and there exists a point with abscissa -1 and a point with abscissa 1 such that if these two points are added to f then the extended simple graph with X-projection $[-1, 1]$ is continuous at each of its points.

Note 2. If h is in $M(-1, 1)$, f is in $M_1(-1, 1)$, and $Zf = h$, then assuming that f' is bounded on the segment Δ, if c is a positive number, there exists a positive number δ such that if $[u, v]$ is an interval with

$$-1 < u < 1 + \delta, \quad \text{and} \quad 1 - \delta < v < 1,$$

then $| \int_u^v h | < c$. That is, the extended integral

$$\int_{-1}^1 h = 0.$$

Note 3. If h is in $M(-1, 1)$, f is in $M_1(-1, 1)$, and $Zf = h$, then $f + \underline{1}$ is in $M_1 (-1, 1)$ and $Z(f + \underline{1}) = h$. This suggests that we impose on f some such condition as $\int_{-1}^1 f = 0$.

Suggestion. Denote by G the subspace of $M(-1, 1)$ in which *point* means an h in $M(-1, 1)$ that is bounded on the segment Δ such that $\int_{-1}^1 h = 0$, and by G_1 the subspace of $M_1(-1, 1)$ in which *point* means an f in $M_1(-1, 1)$ such that f' is bounded on Δ and $\int_{-1}^1 f = 0$. Perhaps Z is a reversible transformation from G_1 to G.

Simple Kernels

The statement that K is a *simple kernel* means that K is a simple surface with XY-projection a rectangular interval $[ab; ab]$ that is symmetric, so $K(x, y) = K(y, x)$ for every x and y in $[a, b]$, *continuous*, and, for some point P in $[ab; ab]$, $K(P) \neq 0$. If h is in $M[a, b]$, the simple graph f, defined by

$$f(x) = \int_a^b K[x, I]h, \quad a \leq x \leq b,$$

is in $M[a, b]$. In fact, this equation defines a linear transformation from $M[a, b]$ to a subset of $M[a, b]$. We use K to denote both the simple kernel and this transformation, so

$$f = Kh.$$

If each of u and v is in $M[a, b]$, then

$$((Ku, v)) = ((u, Kv)).$$

Definition. The statement that λ is a *proper value* of the simple kernel K and ϕ is a *proper function* of K belonging to λ means that λ is a number, ϕ is in $M[a, b]$ and is distinct from $\underline{0}$, and

$$\phi = \lambda \cdot K\phi.$$

Theorem. *If λ_1 and λ_2 are proper values of the simple kernel K and ϕ_1 and ϕ_2 are proper functions of K belonging to λ_1 and λ_2, respectively, then $((\phi_1, \phi_2)) = 0$.*

Definition. The statement that the element f of $M[a, b]$ is *orthogonal* to the element g of $M[a, b]$ means that

$$((f, g)) = 0.$$

The set S of elements of $M[a, b]$ is an *orthonormal* set in case each f in S has norm 1 and each two elements of S are orthogonal to one another.

Exercises.

1. Show that if $K(x, y) = xy$ for every point (x, y) in $[01; 01]$, then K has just one proper value.

2. Show that if $K(x, y) = x + y$ for every point (x, y) in $[01; 01]$, then K has just two proper values.

3. If K is the simple kernel whose XY-projection is $[01; 01]$, defined by

$$K(x, y) = \begin{cases} x(1 - y), & 0 \leq x \leq y \leq 1, \\ y(1 - x), & 0 \leq y \leq x \leq 1, \end{cases}$$

then λ is a proper value of K and ϕ a proper function of K belonging to λ only if λ is a number and ϕ an element of $M[0, 1]$ such that $\phi \neq \underline{0}$, $\phi(0) = 0$, $\phi(1) = 0$, and $\phi'' = -\lambda \cdot \phi$. Hence show that the proper values of K are

$$\lambda_p = \pi^2 p^2, \quad p = 1, 2, 3, \ldots,$$

and that the simple graph ϕ is a proper function of K belonging to λ_p only if ϕ is a product $k \cdot \phi_p$, where k is a number distinct from 0 and

$$\phi_p = \sqrt{2} \cdot S[\pi p I], \quad p = 1, 2, 3, \ldots.$$

The factor $\sqrt{2}$ is introduced so that $\|\phi_p\| = 1$. Thus, $\phi_p, p = 1, 2, 3, \ldots$, is an orthonormal set.

4. Suppose $Vf = -f''$ and c is the pair of endpoint conditions

$$f(0) = -f(1), \quad f'(0) = -f'(1).$$

Show that the kernel of V under c is

$$K(x, y) = \begin{cases} \dfrac{1}{2}(x - y) + \dfrac{1}{4}, & 0 \leq x \leq y \leq 1, \\ \dfrac{1}{2}(y - x) + \dfrac{1}{4}, & 0 \leq y \leq x \leq 1. \end{cases}$$

The proper values of K are

$$\lambda_p = (2p - 1)^2 \pi^2, \quad p = 1, 2, \ldots,$$

and ϕ is a proper function of K belonging to λ_p only if there exists an ordered number pair (k_1, k_2) such that $k_1 \neq 0$ or $k_2 \neq 0$ and $\phi = k_1 C[(2p - 1)\pi I] + k_2 S[(2p - 1)\pi I]$.

5. Suppose $\phi_1, \ldots, \phi_n$ is a finite orthonormal subset of $M[a, b]$ and each of $\lambda_1, \ldots, \lambda_n$ is a number distinct from 0. If K is the simple graph in space with XY-projection $[ab; ab]$ defined by

$$K(x, y) = \sum_{p=1}^{n} \frac{\phi_p(x)\phi_p(y)}{\lambda_p}, \quad a \leq \frac{x}{y} \leq b,$$

then

(i) K is a simple kernel,

(ii) λ is a proper value of K only if there is a positive integer p not greater than n such that $\lambda = \lambda_p$,

(iii) ϕ is a proper function of K belonging to the proper value λ of K only if there exists an ordered number n-tuple $(k_1, \ldots, k_2)$, such that $|k_1| + \cdots + |k_n| > 0$, $\phi = k_1\phi_1 + \cdots + k_n\phi_n$, and $k_p = 0$ for each positive integer p not greater than n such that $\lambda_p \neq \lambda$.

Theorem. *If K is a simple kernel, then K has a proper value.*

The reader may assume the theorem is true and use it as an axiom if needed. The symmetry of K is essential: if $K(x, y) = S(x) \cdot C(y)$ and $[a, b]$ is $[0, 2\pi]$, there is no proper value.

An Identity and Some Consequences

Suppose $\phi_1, \ldots, \phi_n$ is a finite orthonormal subset of $M[a, b]$, f is a point of $M[a, b]$, and $(k_1, \ldots, k_n)$ is a number n-tuple. Then

$$\left\| f - \sum_{p=1}^{n} k_p \cdot \phi_p \right\|^2 = \|f\|^2 - \sum_{p=1}^{n}((f, \phi_p))^2 + \sum_{p=1}^{n}\{((f, \phi_p)) - k_p\}^2.$$

From this identity there follows

i. $\left\| f - \sum_{p=1}^{n} k_p \cdot \phi_p \right\| > \left\| f - \sum_{p=1}^{n} ((f, \phi_p)) \cdot \phi_p \right\|$ unless $k_p = ((f, \phi_p))$, $p = 1, \ldots, n$,

ii. $\sum_{p=1}^{n} ((f, \phi_p))^2 \leq |f|^2$, and

iii. if ϕ_p is a proper function of the simple kernel K belonging to the proper value λ_p, $p = 1, \ldots, n$, then

$$\sum_{p=1}^{n} \left(\frac{1}{\lambda_p} \right)^2 \leq \int_a^b \int_a^b \{K(x, y)\}^2 dx \, dy.$$

Consequently,

(a) no interval includes an infinite set of proper values of K,

(b) if m of the ϕ_p belong to the same proper value λ of K, then $m \leq \lambda^2 \int_a^b \int_a^b \{K(x, y)\} dx \, dy$, and

(c) if the set of proper values of the simple kernel K is an infinite set M, there exists a number sequence $\{\lambda_p\}_{p=1}^{\infty}$ such that λ is a proper value of K only if there is a positive integer p such that $\lambda = \lambda_p$ and $|\lambda_p| \leq |\lambda_{p+1}|$ for every positive integer p.

Some Geometry in the Space $M[a, b]$

Suppose $f_1, \ldots, f_n$ are n points of $M[a, b]$ such that if $(k_1, \ldots, k_n)$ is a number n-tuple and $k_1 \cdot f_1 + \cdots + k_n \cdot f_n = \underline{0}$, then $k_1 = \cdots = k_n = 0$. Denote by R the point set in $H[a, b]$ to which f belongs only if there exists a number n-tuple $(k_1, \ldots, k_n)$ such that $f = k_1 \cdot f_1 + \cdots + k_n \cdot f_n$. R is the point set of a linear subspace of $M[a, b]$ that may be called an n-plane in $M[a, b]$.

Theorem. *If g is a point of $M[a, b]$, then there exists a point f in the n-plane R which is nearer g than any other point of R, so*

$$\|g - f\| < \left\| g - \sum_{p=1}^{n} k_p \cdot f_p \right\|$$

unless $f = \sum_{p=1}^{n} k_p \cdot f_p$.

Suggestion. Settle this first in case $f_1, \ldots, f_n$ is an orthonormal set.

Complete Orthonormal Sets

Definition. The statement that the subset S of $M[a, b]$ is a *complete orthonormal set* of proper functions of the simple kernel K means that each f in S is a proper function of K, that S is an orthonormal set, and that if ϕ is a proper function of K then there exists a number n-tuple $(k_1, \ldots, k_n)$ such that

$$\phi = \sum_{p=1}^{n} k_p \cdot \phi_p.$$

Theorem. *If K is a simple kernel, then K has a complete orthonormal set S of proper functions that is a finite set or an infinite set that is an infinite sequence $\{\phi_p\}_{p=1}^{\infty}$ such that ϕ belongs to S only if there exists a positive integer p for which $\phi = \phi_p$.*

The First Expansion Theorem

Assume for the present that every simple kernel has a proper value. It follows from this assumption that, if G is a simple surface whose XY-projection is the rectangular interval $[ab; ab]$ that is symmetric, continuous at each of its ordered pairs, and has no proper value, then $G(P) = 0$ for every point P in $[ab; ab]$.

Theorems. *Suppose K is a simple kernel.*

1. *Suppose the finite orthonormal subset $\phi_1, \ldots, \phi_n$ of $M[a, b]$ is a complete orthonormal set of proper functions of K and ϕ_p belongs to the proper value λ_p of K, $p = 1, \ldots, n$. If G is the simple surface with XY-projection $[ab; ab]$, defined for each point (x, y) in $[ab; ab]$ by*

$$G(x, y) = K(x, y) - \sum_{p=1}^{n} \frac{\phi_p(x)\phi_p(y)}{\lambda_p},$$

then G is continuous and symmetric and has no proper value, so that $G(x, y) = 0$ for every (x, y) in $[ab; ab]$. That is,

$$K(x, y) = \sum_{p=1}^{n} \frac{\phi_p(x)\phi_p(y)}{\lambda_p}, \quad a \le \frac{x}{y} \le b.$$

Moreover, if the simple graph f is expressible in terms of a simple graph h in $M[a,b]$ as

$$f = Kh,$$

then $f(x) = \sum_{p=1}^{n}((f, \phi_p)) \cdot \phi_p(x), a \le x \le b$.

2. *Suppose $\{\phi_p\}_{p=1}^{\infty}$ is a complete orthonormal set of proper functions of K and ϕ_p belongs to the proper value λ_p of K, $p = 1, 2, 3, \ldots$. Suppose also that there exists a simple surface k with XY-projection $[ab; ab]$ such that if c is a positive number then there exists a positive integer N such that if n is an integer greater than N and (x, y) is in $[ab; ab]$, then*

$$\left| k(x, y) - \sum_{p=1}^{n} \frac{\phi_p(x)\phi_p(y)}{\lambda_p} \right| < c.$$

It is true that
(i) k is continuous and symmetric,

(ii) if $G = k - K$, then G is continuous and symmetric and has no proper value, so that $G(x, y) = 0$ for every (x, y) in $[ab; ab]$, or

$$k = K.$$

We express this last result briefly as

$$K(x, y) = \sum_{p=1}^{\infty} \frac{\phi_p(x)\phi_p(y)}{\lambda_p} \quad \text{uniformly on } [ab; ab].$$

Analogous abbreviations will be used in the sequel.

Also, if the simple graph f is expressible in terms of a simple graph h in $M[a, b]$ as $f = Kh$, then

$$f = \sum_{p=1}^{\infty}((f, \phi_p)) \cdot \phi_p \quad \text{uniformly on } [a, b].$$

The last statement means that if c is a positive number then there exists a positive integer N such that if n is an integer greater than N and x is in $[a, b]$ then

$$\left| f(x) - \sum_{p=1}^{n}((f, \phi_p)) \cdot \phi_p(x) \right| < c.$$

An Important Special Case

We have seen that the set $\phi_p = \sqrt{2} \cdot S[\pi p I]$, $p = 1, 2, 3, \ldots$, is a complete orthonormal set of proper functions of the simple kernel K defined by

$$K(x, y) = \begin{cases} x(1 - y), & 0 \le x \le y \le 1, \\ y(1 - x), & 0 \le y \le x \le 1. \end{cases}$$

The hypothesis of Theorem 2 of the last section is satisfied, and so

(i) $$K(x, y) = \sum_{p=1}^{\infty} \frac{2 \cdot S(\pi p x) S(\pi p y)}{\pi^2 p^2}$$

uniformly for $0 \le \dfrac{x}{y} \le 1$.

The simple graph f is expressible in terms of an h of $M[0, 1]$ only if f'' has X-projection $[0, 1]$ f'' is continuous, and $f(0) = 0$, $f(1) = 0$. For any such f

(ii) $$f(x) = \sum_{p=1}^{\infty} 2 \cdot \left\{ \int_0^1 f \cdot S[\pi p I] \right\} \cdot S(\pi p x)$$

uniformly for $0 \le x \le 1$.

There are interesting special cases. For instance, if $y = x$, (i) becomes

(iii) $$x(1 - x) = \sum_{p=1}^{\infty} \frac{2 S^2(\pi p x)}{\pi^2 p^2}.$$

Exercises.

i. The slope, $D_x\{I(1 - I)\}$, is $1 - 2x$. It is obviously *not* true that

$$1 - 2x = \sum_{p=1}^{\infty} D_x \left\{ \frac{2 S^2[p\pi I]}{\pi^2 p^2} \right\} = 2 \sum_{p=1}^{\infty} \frac{S(2\pi p x)}{\pi p}$$

if $x = 0$ or if $x = 1$. Prove that this formula *is* true *uniformly* for $u \le x \le v$ for any interval $[u, v]$ such that $0 < u < v < 1$.

ii. Show that $\frac{\pi^2}{8} = \sum_{p=1}^{\infty} \left\{ \frac{1}{2p-1} \right\}^2$.

iii. Investigate the simple graphs F_k defined by

$$F_1(x) = \begin{cases} 0, & x = 0, \\ x - (1/2), & 0 < x < 1, \\ 0, & x = 1, \end{cases} \quad F_k(x) = F_k(x+1); \; \int_0^1 F_k = 0;$$

$$F_{k+1}(x) = F_{k+1}(0) + \int_0^x F_k, \quad \text{if } 0 \le x \le 1.$$

Second Expansion Theorem

Definition. The statement that the simple kernel K is *positive definite* means that if h is in $M[a, b]$ then

$$\int_a^b \int_a^b K(x, y) h(x) h(y) dx \, dy \ge 0.$$

Theorem. *If the simple kernel K is positive definite, then every proper value of K is positive.*

Suppose $\phi_1, \ldots, \phi_n$ is an orthonormal subset of $M[a, b]$ and each of $\lambda_1, \ldots, \lambda_n$ is a number distinct from 0 and consider the simple kernel (studied previously),

$$K(x, y) = \sum_{p=1}^n \frac{\phi_p(x)\phi_p(y)}{\lambda_p}.$$

Is there a simple formula, involving an integral, relating K to the simple kernel

$$K^{(2)}(x, y) = \sum_{p=1}^n \frac{\phi_p(x)\phi_p(y)}{\lambda_p^2} \; ?$$

Certainly every proper value of $K^{(2)}$ is positive.

Definition. If K is a simple kernel and n is a positive integer, $K^{(n)}$ (read K *upper n*) is defined for each point (x, y) in $[ab; ab]$ by

$$K^{(1)} = K,$$

$$K^{(n+1)}(x, y) = \int_a^b K^{(n)}[x, I] \cdot K[I, y].$$

Theorems. *Suppose K is a simple kernel.*

1. *If each of m and n is a positive integer,*

$$K^{(m+n)}(x, y) = \int_a^b K^{(m)}[x, I] \cdot K^{(n)}[I, y].$$

2. *If n is a positive integer, then $K^{(2n)}$ is a positive definite simple kernel.*

3. *If the simple kernel K has the proper value λ, then $K^{(2)}$ has the proper value λ^2. If $K^{(2)}$ has the proper value λ^2, then K has the proper value λ or K has the proper value $-\lambda$.*

4. *Any complete orthonormal set of proper functions of K is a complete orthonormal set of proper functions of $K^{(2)}$ (hence, of $K^{(4)}$, $K^{(8)}$, ...).*

Lemma. *If each of $a_1, \ldots, a_n$ and $b_1, \ldots, b_n$ is a number n-tuple, then*

$$\left\{ \sum_{p=1}^n a_p b_p \right\}^2 \le \left\{ \sum_{p=1}^n a_p^2 \right\} \left\{ \sum_{p=1}^n b_p^2 \right\}.$$

Theorems. *Suppose K is a simple kernel and $\{\phi_p\}_{p=1}^\infty$ is a complete orthonormal set of proper functions of K and ϕ_p belongs to the proper value λ_p of K, $p = 1, 2, 3, \ldots$.*

1. *There exists a number d^2 such that if t is in $[a, b]$ then*

$$\sum_{p=1}^n \left[\frac{\phi_p(t)}{\lambda_p} \right]^2 \le d^2, \quad n = 1, 2, 3, \ldots.$$

2. *There exists a simple surface k with XY-projection $[ab; ab]$ such that if x is in $[a, b]$ then*

$$k(x, y) = \sum_{p=1}^\infty \frac{\phi_p(x)\phi_p(y)}{\lambda_p^2}$$

uniformly for $a \le y \le b$.

3.

$$K^{(4)}(x, y) = \sum_{p=1}^\infty \frac{\phi_p(x)\phi_p(y)}{\lambda_p^4}$$

uniformly for $a \le \dfrac{x}{y} \le b$.

4. *$k = K^{(2)}$.*

5. *If h is in $M[a, b]$ and $\int_a^b \phi_p h = 0$, so $((\phi_p, h)) = 0$ for every positive integer p, then if x is in $[a, b]$*

$$\int_a^b K^{(4)}[x, I]h = 0,$$

$$\int_a^b K^{(2)}[x, I]h = 0,$$

$$\int_a^b K[x, I]h = 0.$$

6. *If the simple graph f can be represented in terms of a g in $M[a, b]$ by $f = Kg$, then*

$$f = \sum_{p=1}^{\infty} ((f, \phi_p)) \cdot \phi_p \text{ uniformly on } [a, b].$$

7. *If g is in $M[a, b]$,*

$$\int_a^b \int_a^b K(x, y)g(x)g(y)dx\, dy = \sum_{p=1}^{\infty} \frac{((g, \phi_p))^2}{\lambda_p}.$$

8. *If every proper value of K is positive, then K is positive definite. (By this and an earlier result, K is positive definite only if each of its proper values is positive.)*

9. *If K has a positive proper value and λ_1 is the smallest positive proper value of K then, if g is in $M[a, b]$ and $\|g\| = 1$,*

$$\int_a^b \int_a^b K(x, y)g(x)g(y)dx\, dy \leq \frac{1}{\lambda_1},$$

with equality if $g = \phi_1$.

Third Expansion Theorem

Lemma. *Suppose that for each positive integer n, f_n belongs to $M[a, b]$ and if x is in $[a, b]$, $f_n(x) \leq f_{n+1}(x)$. Suppose f belongs to $M[a, b]$ and if x is in $[a, b]$ and c is a positive number then there is a number N such that if $n > N, 0 \leq f(x) - f_n(x) < c$. Then, if c is a positive number there exists a number N such that if x is in $[a, b]$ and $n > N$, then $0 \leq f(x) - f_n(x) < c$.*

Theorems. *Suppose K is a positive definite simple kernel with the complete orthonormal set of proper functions $(\phi_p)_{p=1}^{\infty}$ and suppose that, for each positive integer p, ϕ_p belongs to the proper value λ_p of K.*

1. *If for each positive integer n, K_n is the simple surface defined for each (x, y) in $[ab; ab]$ by*

$$K_n(x, y) = K(x, y) - \sum_{p=1}^{n} \frac{\phi_p(x)\phi_p(y)}{\lambda_p},$$

   then K_n is a positive definite simple kernel.

2. *If x is in $[a, b]$ and n is a positive integer, $K_n(x, x) \geq 0$, so*

$$\sum_{p=1}^{n} \frac{\phi_p^2(x)}{\lambda_p} \leq K(x, x).$$

3. *There exists a simple surface k with XY-projection $[ab; ab]$ such that if x is in $[a, b]$*

$$k(x, y) = \sum_{p=1}^{\infty} \frac{\phi_p(x)\phi_p(y)}{\lambda_p}$$

   uniformly for $a \leq y \leq b$.

4. *The simple graph g defined for each x in $[a, b]$ by $g(x) = k(x, x)$ is continuous.*

5.

$$g(x) = \sum_{p=1}^{\infty} \frac{\phi_p^2(x)}{\lambda_p} \quad \text{uniformly on } [a, b].$$

6.

$$K(x, y) = \sum_{p=1}^{\infty} \frac{\phi_p(x)\phi_p(y)}{\lambda_p} \quad \text{uniformly on } [ab; ab].$$

Corollary. *If K is a simple kernel, then*

$$K^{(2)}(x, y) = \sum_{p=1}^{\infty} \frac{\phi_p(x)\phi_p(y)}{\lambda_p^2} \quad \text{uniformly on } [ab; ab].$$

Every Simple Kernel has a Proper Value

A study of the simple kernel K, defined in terms of the orthonormal subset $\phi_1, \ldots, \phi_n$ of $M[a, b]$ and the number n-tuple $\lambda_1, \ldots, \lambda_n$ by

$$K(x, y) = \sum_{p=1}^{n} \frac{\phi_p(x)\phi_p(y)}{\lambda_p},$$

furnishes a clue to a proof that every simple kernel has a proper value.

If m is a positive integer

$$K^{(m)}(x, y) = \sum_{p=1}^{n} \frac{\phi_p(x)\phi_p(y)}{\lambda_p^m},$$

so

$$\int_a^b K^{(m)}[I, I] = \sum_{p=1}^{n} \left(\frac{1}{\lambda_p}\right)^m.$$

If λ^2 denotes the largest number γ such that $\gamma \leq (\lambda_p)^2$, $p = 1, \ldots, n$, and $u_m = \sum_{p=1}^{n} \left(\frac{1}{\lambda_p}\right)^{2m}$, $m = 1, 2, 3, \ldots$, then it is easy to see that the number sequence $\left\{\frac{u_m}{u_{m+1}}\right\}_{m=1}^{\infty}$ converges to the sequential limit λ^2. This suggests that for any simple kernel K if

$$u_m = \int_a^b K^{(2m)}[I, I], \qquad m = 1, 2, 3, \ldots,$$

then the number sequence $\left\{\frac{u_m}{u_{m+1}}\right\}_{m=1}^{\infty}$ may be proved to converge to λ^2, that λ^2 is a proper value of $K^{(2)}$, and so K has the proper value λ or else the proper value $-\lambda$.

Orthogonal Polynomials

We return to the transformation Z of an earlier section defined by

$$Zf = -\{(1 - I^2)f'\}',$$

for f with X-projection the segment Δ with ends -1 and 1, having a continuous second derivative f'' with X-projection Δ.

G_1 denotes the set of all such simple graphs f such that f', and therefore f, is bounded on Δ.

Definition. The statement that λ is a proper value of the transformation Z and ϕ a proper function of Z belonging to λ means that λ is a number and ϕ an element of G_1 distinct from $\underline{0}$ such that

$$Z\phi = \lambda \cdot \phi.$$

Examples.

i. 0 is a proper value of Z and $\underline{1}$ (restricted to Δ) a proper function of Z belonging to the proper value 0.

ii. 2 is a proper value of Z and I (restricted to Δ) a proper function of Z belonging to the proper value 2.

We restrict all simple graphs in this discussion to the segment Δ.

If f is a simple graph, $f^{(0)}$, the zeroth derivative of f denotes f itself, $f^{(1)} = f'$, $f^{(2)} = f''$, and if n is a positive integer, $f^{(n+1)} = \{f^{(n)}\}'$; $f^{(n)}$ is called the nth derivative of f.

Lemma. *If n is a positive integer,*

$$(uv)^{(n)} = u^{(n)}v^{(0)} + \frac{n}{1}u^{(n-1)}v^{(1)} + \frac{n(n-1)}{1\cdot 2}u^{(n-2)}v^{(2)} + \cdots$$
$$+ \frac{n(n-1)\cdots(n-r+1)}{1\cdot 2\cdot\cdots\cdot r}u^{(n-r)}v^{(r)} + \cdots + u^{(0)}v^{(n)}$$

for any simple graph u and simple graph v having the derivatives indicated.

If n is a positive integer, the statement that Q_n is a *polynomial of degree n* means that Q_n is a simple graph $a_0 I^n + a_1 I^{n-1} + \cdots + a_n$ where each of $a_0, \ldots, a_n$ is a number and $a_0 \neq 0$.

Theorems. *Suppose λ is a number, n is a positive integer, and Q_n is a polynomial of degree n such that $Z Q_n = \lambda \cdot Q_n$.*

1. If k is a positive integer,

$$\left\{(1 - I^2)^k Q_n^{(k)}\right\}' = -\{\lambda - (k-1)k\}(1 - I^2)^{k-1} Q_n^{(k-1)}.$$

2. $\lambda = n(n+1)$.

3.

$$\{(1 - I^2)^n Q_n^{(n)}\}^{(n)} =$$
$$(-1)^n\{\lambda - (n-1)n\}\{\lambda - (n-2)(n-1)\}\cdots\{\lambda - 1\cdot 2\}\cdot\lambda\cdot Q_n,$$

so that for some number k_n distinct from 0,

$$Q_n = k_n \cdot \{(1 - I^2)^n\}^{(n)}.$$

Examples. $Q_0 = k_0$, $Q_1 = k_1 \cdot (-2I)$, and $Q_2 = k_2 \cdot \{-4(1 - 3I^2)\}$. We determine $k_0, k_1, \ldots$ so that $Q_n(1) = 1$. Then $k_0 = 1$, $k_1 = -(1/2)$, and $k_2 = (1/8)$. We denote the polynomial Q_n with k_n so determined by P_n. Then

$$P_0 = \underline{1}, \quad P_1 = I, \quad P_2 = \frac{1}{2}(3I^2 - 1).$$

Theorems. *Suppose n is a positive integer and*

$$Q_n = \{(1 - I^2)^n\}^{(n)}.$$

1. $Q_n = (1 - I^2)Q'_{n-1} - 2nI\,Q_{n-1} - n(n-1)h$, *where* $h(x) = \int_{-1}^{x} Q_{n-1}$.

2. *If* $Z Q_{n-1} = (n-1)n\,Q_{n-1}$, *then*

$$Q'_n = -2nI\,Q'_{n-1} - 2n^2 Q_{n-1},$$
$$Q_n = 2(1 - I^2)Q'_{n-1} - 2nI\,Q_{n-1}, \quad and$$
$$Z Q_n = n(n+1)Q_n.$$

3. $Q_n(1) = (-1)^n 2^n \cdot 1 \cdot 2 \cdots \cdots n.$

4. $P_0 = \underline{1}$ *and* $P_n = \dfrac{1}{2^n \cdot 1 \cdot 2 \cdots \cdots n} \cdot \{(I^2 - 1)^n\}^{(n)}.$

Since $Z P_n = n(n+1)P_n$, $n = 0, 1, 2, \ldots$, $n(n+1)$ is a proper value of Z and P_n is a proper function of Z belonging to the proper value $n(n+1)$ of Z.

Theorem. *If ϕ is a proper function of Z belonging to the proper value $n(n+1)$ where n is a nonnegative integer, then there exists a number k distinct from 0 such that $\phi = k \cdot P_n$. If λ is a number distinct from $n(n+1)$, $n = 0, 1, 2, \ldots$, then λ is not a proper value of Z.*

Theorem. *If each of m and n is a nonnegative integer,*

$$\int_{-1}^{1} P_m P_n = \begin{cases} 0, & m \neq n, \\ \frac{2}{2n+1}, & m = n. \end{cases}$$

14

More about Linear Spaces

The space in which *point* means number sequence, with addition of the number sequence a and the number sequence b defined by

$$(a + b)_p = a_p + b_p, \quad p = 1, 2, 3, \ldots,$$

and multiplication of a by the number k defined by

$$(k \cdot a)_p = ka_p, \quad p = 1, 2, 3, \ldots,$$

is a linear space. We denote this linear space of *all* number sequences by W. If a is a point of W for which there is a number k_1 such that

$$\sum_{p=1}^{n} a_p^2 \leq k_1, \quad n = 1, 2, 3, \ldots,$$

and b is a point of W for which there is a number k_2 such that

$$\sum_{p=1}^{n} b_p^2 \leq k_2, \quad n = 1, 2, 3, \ldots,$$

then

$$\sum_{p=1}^{n}(a_p + b_p)^2 = \sum_{p=1}^{n} a_p^2 + \sum_{p=1}^{n} b_p^2 + 2 \sum_{p=1}^{n} a_p b_p$$

$$\leq k_1 + k_2 + 2 \left\{ \sum_{p=1}^{n} a_p^2 \cdot \sum_{p=1}^{n} b_p^2 \right\}^{1/2} \leq \left(k_1^{1/2} + k_2^{1/2} \right)^2,$$

so that $a + b$ has the same property. Also, if k is a number, $\sum_{p=1}^{n} (ka)_p^2 = k^2 \sum_{p=1}^{n} a_p^2 \leq k^2 k_1$, so that $k \cdot a$ has the same property.

Definition. $\mathcal{H}$ denotes the linear space in which *point* means number sequence a such that for some number k, $\sum_{p=1}^{n} a_p^2 \leq k$, for every positive integer n, with addition and multiplication by a number defined as in the linear space W and with the inner product defined for every point a and b by

$$((a, b)) = \sum_{p=1}^{\infty} a_p b_p.$$

(That is, if ε is a positive number then there exists a number m such that $|((a, b)) - \sum_{p=1}^{n} a_p b_p| < \varepsilon$ if $n > m$.)

We are interested in the linear space $\mathcal{H}$ primarily because we have seen in the preceding chapter that certain simple graphs f have associated with them points a of $\mathcal{H}$ defined by $a_p = ((f, \phi_p))$, $p = 1, 2, 3, \ldots$. (See the second expansion theorem.)

Properties of the Space $\mathcal{H}$

1. If n is a positive integer, E_n denotes the linear space in which *point* means ordered number n-tuple $(a_1, \ldots, a_n)$ with addition and multiplication by a number k defined by

$$(a_1, \ldots, a_n) + (b_1, \ldots, b_n) = (a_1 + b_1, \ldots, a_n + b_n)$$

and

$$k \cdot (a_1, \ldots, a_n) = (ka_1, \ldots, ka_n)$$

and with inner product $((a, b)) = a_1 b_1 + \cdots + a_n b_n$. E_n is described as *isometric with* the subspace $\mathcal{H}^{(n)}$ of all points a of $\mathcal{H}$ such that $a_p = 0$ if $p > n$. This means that there exists a reversible transformation F from E_n to $\mathcal{H}^{(n)}$ that is *linear*, that is, if a is in E_n, b is in E_n, and k is a number, then $F(a + b) = F(a) + F(b)$ and $F(k \cdot a) = k \cdot F(a)$; and *distance preserving*, that is, the distance from the point a of E_n to the point b of E_n is the distance from the point $F(a)$ of $\mathcal{H}^{(n)}$ to the point $F(b)$ of $\mathcal{H}^{(n)}$. In fact, such a transformation is given by

$$F(a_1, \ldots, a_n) = (a_1, \ldots, a_n, 0, 0, \ldots).$$

2. Suppose G is an infinite point set in E_n that is *bounded*, that is, there is a number r such that if x is in G then $||x|| \leq r$. Then there is a point a of G, called a *limit point* of G, such that if r is a positive number then there exists an infinite subset G_1 of G such that if x is in G_1, $||x - a|| < r$. The space $\mathcal{H}$ does not have this property. There exists an infinite point set G in $\mathcal{H}$ such that if x is in G then $||x|| = 1$, and if x any y are points of G, $||x - y|| > 1$.

3. There exists in $\mathcal{H}$ a point sequence $\{a^{(p)}\}_{p=1}^{\infty}$, that is, if p is a positive integer then $a^{(p)}$ is a point of $\mathcal{H}$ such that if x is a point of $\mathcal{H}$ and r is a positive number then there is a positive integer p such that $||x - a^{(p)}|| < r$.

4. The statement that the point sequence $\{b^{(p)}\}_{p=1}^{\infty}$ in $\mathcal{H}$ is a *closing sequence* means that if ε is a positive number then there exists a number k such that $||b^{(m)} - b^{(n)}|| < \varepsilon$ if $m > k$ and $n > k$. Every closing point sequence in $\mathcal{H}$ converges to a point of $\mathcal{H}$. That is, if $\{b^{(p)}\}_{p=1}^{\infty}$ is a closing point sequence in $\mathcal{H}$, then there exists a point a in $\mathcal{H}$ such that, if ε is a positive number, there exists a number m such that $||a - b^{(n)}|| < \varepsilon$ if $n > m$.

5. There exists a point sequence $\{\phi^{(p)}\}_{p=1}^{\infty}$ in $\mathcal{H}$ so that

(i) $$\left((\phi^{(p)}, \phi^{(q)}) \right) = \begin{cases} 0, & p \neq q, \\ 1, & p = q; \end{cases}$$

and

(ii) if x is a point of $\mathcal{H}$ and ε is a positive number, there exists a number m such that $||x - \sum_{p=1}^{n} ((\phi^{(p)}, x)) \cdot \phi^{(p)}|| < \varepsilon$ if $n > m$.

6. Any inner product space having the properties of $\mathcal{H}$ indicated in 2, 3, and 4 is isometric with $\mathcal{H}$.

Further discussion is undesirable because we do not want to spoil these good problems for the reader.

Linear Transformations in $\mathcal{H}$

Suppose T is a transformation from $\mathcal{H}$ to a number set.

i. The statement that T is *linear* means that if each of a and b is a point of $\mathcal{H}$ and k is a number, then

$$T(a + b) = Ta + Tb \quad \text{and} \quad T(k \cdot a) = kTa.$$

ii. The statement that T is *continuous* means that if a is a point of H and ε is a positive number then there exists a positive number δ such that if x is a point of $\mathcal{H}$ and $||x - a|| < \delta$, then $|Tx - Ta| < \varepsilon$.

iii. The statement that T is *bounded* means there exists a number k such that if x is a point of $\mathcal{H}$, then $|Tx| \le k||x||$.

Theorem. *Suppose T is a linear transformation from $\mathcal{H}$ to a number set. The following two statements are equivalent:*

  (i) *T is continuous*

  and

  (ii) *T is bounded.*

Example. If a is a point of $\mathcal{H}$ and T the transformation from $\mathcal{H}$ to a number set defined for each x in $\mathcal{H}$ by

$$Tx = ((a, x))$$

then T is linear and bounded, and the least number k such that $|Tx| \le k||x||$ for every x in $\mathcal{H}$ is the number $||a||$.

Theorem. *If T is a bounded linear transformation from $\mathcal{H}$ to a number set, then there exists a point a of $\mathcal{H}$ such that if x is a point of $\mathcal{H}$ then $Tx = ((a, x))$.*

Problem. Show that there is a linear space isometric with $\mathcal{H}$ in which point means bounded linear transformation from $\mathcal{H}$ to a number set.

1. Suppose F is a transformation from the set of ordered pairs (x, y), each member of which is a point of $\mathcal{H}$, to a number set.

  i. The statement that F is *bilinear* means that if each of x, y, and z is a point of $\mathcal{H}$ and k is a number then

  $$F(x + y, z) = F(x, z) + F(y, z),$$
  $$F(x, y + z) = F(x, y) + F(x, z),$$

  and

  $$F(k \cdot x, y) = F(x, k \cdot y) + kF(x, y).$$

  ii. The statement that F is *bounded* means that there exists a number k such that if each of x and y is a point of $\mathcal{H}$ then

  $$|F(x, y)| \le k\,||x||\,||y||.$$

2. Suppose f is a transformation from the set of points of $\mathcal{H}$ to a point set in $\mathcal{H}$.

   i. The statement that f is *linear* means that if each of x and y is a point of $\mathcal{H}$ and k is a number then

$$f(x + y) = fx + fy \quad \text{and} \quad f(k \cdot x) = k \cdot fx.$$

   ii. The statement that f is *bounded* means that there exists a number k such that if x is a point of $\mathcal{H}$ then

$$\|fx\| \leq k\|x\|.$$

Theorems. *Suppose F is a transformation from the ordered pairs (x, y) with first member a point of $\mathcal{H}$ and second member a point of $\mathcal{H}$ to a number set.*

1. Suppose that f is a bounded linear transformation from $\mathcal{H}$ to a point set in $\mathcal{H}$. If $F(x, y) = ((fx, y))$, where each of x and y is a point of $\mathcal{H}$, then F is bilinear and bounded.

2. If F is bilinear and bounded, there exists only one bounded linear transformation f from $\mathcal{H}$ to a point set in $\mathcal{H}$ and only one bounded linear transformation f^* from $\mathcal{H}$ to a point set in $\mathcal{H}$ such that if x is a point of $\mathcal{H}$ and y a point of $\mathcal{H}$ then

$$F(x, y) = ((fx, y)) \quad \text{and} \quad F(x, y) = ((x, f^*y)).$$

The bounded linear transformations f and f^* from $\mathcal{H}$ to point sets in $\mathcal{H}$ are called adjoints *of one another.*

Definition. The statement that f is a *matrix* means that f is a transformation from the set of ordered positive integer pairs to a number set. If f is a matrix and (p, q) is an ordered number pair, each member of which is a positive integer, we denote $f(p, q)$ by f_{pq}. The statement that the matrix f is *bounded* means that there exists a number k such that if n is a positive integer and each of $x_1, \cdots, x_n$ and $y_1, \cdots y_n$ is a number n-tuple, then

$$\left| \sum_{p,q=1}^{n} x_p f_{pq} y_q \right| \leq k \left\{ \sum_{p=1}^{n} x_p^2 \cdot \sum_{q=1}^{n} y_q^2 \right\}^{1/2}.$$

Theorem. *The set of bounded matrices may be identified with the set of bounded linear transformations from $\mathcal{H}$ to point sets in $\mathcal{H}$ so that if f is such a transformation, if x is the point $\{x_p\}_{p=1}^{\infty}$ of $\mathcal{H}$, if $fx = y = \{y_p\}_{p=1}^{\infty}$, and if the same letter f is used to denote the bounded matrix identified with* *this transformation, then*

$$y_p = \sum_{q=1}^{\infty} f_{pq} x_q, \quad p = 1, 2, 3, \ldots.$$

The adjoint f^* of f is defined by

$$f_{pq}^* = f_{qp}.$$

The Space $C_0[a, b]$

This is the linear space in which *point* means simple graph f with the properties that f has X-projection the interval $[a, b]$, f is continuous, $f(a) = 0$, and the norm of f is defined by

$$\|f\|_0 = \left[\begin{array}{c} \text{The least number k such that} \\ |f(x)| \le k \text{ for every } x \text{ in } [a, b] \end{array} \right].$$

Suppose $\{t_p\}_{p=1}^{\infty}$ is a number sequence such that

i. $a < t_p \le b, \ p = 1, 2, 3, \ldots,$

ii. $t_p \ne t_q$, if $p \ne q$,

and

iii. if s is a subinterval of $[a, b]$, there exists a positive integer p such that t_p is in s.

For each positive integer p, θ_p denotes the simple graph in $C_0[a, b]$ defined by

$$\theta_p(x) = \begin{cases} x, & a \le x \le t_p, \\ t_p, & t_p \le x \le b. \end{cases}$$

Theorems.

1. If n is a positive integer and $k_1, \ldots, k_n$ is a number n-tuple such that $k_1 \cdot \theta_1 + \cdots + k_n \cdot \theta_n = 0$, then $k_1 = \cdots = k_n = 0$.

2. If f is a point of $C_0[a, b]$ and ε a positive number, then there exists a positive integer n and a number n-tuple $k_1, \ldots, k_n$ such that

$$\left\| f - \sum_{p=1}^{\infty} k_p \cdot \theta_p \right\|_0 < \varepsilon.$$

The Space $BV_0[a, b]$

We state a theorem that includes some ideas previously introduced.

Theorem. *Suppose f is a simple graph with X-projection the interval $[a, b]$ that is of bounded variation on $[a, b]$.*

  1. If c is a number between a and b, then f is of bounded variation on $[a, c]$ and on $[c, b]$ and the total variation of f on $[a, b]$ is the sum of the total variation of f on $[a, c]$ plus the total variation of f on $[c, b]$. That is,

$$V_a^b f = V_a^c f + V_c^b f.$$

  2. If ϕ_1 denotes the simple graph with X-projection $[a, b]$ such that

$$\phi_1(x) = \begin{cases} 0, & x = a, \\ V_a^x f & a < x \le b, \end{cases}$$

  then

  (i) ϕ_1 is nondecreasing on $[a, b]$,

  (ii) $f + \phi_1$ is nondecreasing on $[a, b]$,

  and

  (iii) If $f + \phi_1$ is denoted by ϕ_2, we see that $f = \phi_2 - \phi_1$, the difference of two nondecreasing simple graphs.

  3. f has property Q at each of its points.

In the notation in the first of the two questions on page 92, if the abscissa of P is x, we denote P_L by $(x, f(x-))$ if $a < x \le b$ and P_R by $(x, f(x+))$ if $a \le x < b$.

Definition. $BV_0[a, b]$ is the linear space in which *point* means simple graph f with the properties that f has X-projection $[a, b]$, $f(a) = 0$, f is of bounded variation on $[a, b]$, the norm of f is

$$\|f\|_{bv} = V_a^b f,$$

and if $a \le x < b$,

$$f(x) = f(x+).$$

Theorem. *If f is in $BV_0[a, b]$, denote by $\|f\|_0$ (as in the space $C_0[a, b]$) the least number k such that $|f(x)| \le k$ for every x in $[a, b]$. If f is* *in $BV_0[a, b]$, then*

$$\|f\|_0 \leq \|f\|_{bv}.$$

The Space $C_m[a, b]$

The points of this space are the same as the points of $C_0[a, b]$, but there is a different norm. The letter m denotes a simple graph with X-projection $[a, b]$ that is *increasing*; i.e., if $a \leq x < y \leq b$, then $m(x) < m(y)$, $m(a) = 0$, and, if $a \leq x < b$, $m(x) = m(x+)$. If f is in $C_m[a, b]$ and g is in $C_m[a, b]$, the inner product of f and g is

$$((f, g))_m = \int_a^b f g \, dm.$$

Thus, $C_m[a, b]$ is an inner product space with norm

$$\|f\|_m = \left\{ \int_a^b f^2 \, dm \right\}^{1/2}.$$

Theorem. *Suppose $\theta_1, \theta_2, \theta_3, \ldots$ is the sequence of simple graphs in $C_0[a, b]$ defined previously in terms of the number sequence $\{t_p\}_{p=1}^{\infty}$,*

$$\phi_1 = \frac{\theta_1}{\|\theta_1\|_m}$$

and, if p is a positive integer,

$$\phi_{p+1} = \frac{\theta_{p+1} - \sum_{q=1}^{p} ((\theta_{p+1}, \phi_q))_m \cdot \phi_q}{\left\| \theta_{p+1} - \sum_{q=1}^{p} ((\theta_{p+1}, \phi_q))_m \cdot \phi_q \right\|_m}.$$

Then the following statements are true.

(i) $((\phi_i, \phi_j))_m$ is 0 if $i \neq j$ and 1 if $i = j$.

(ii) If n is a positive integer, there exists a number n-tuple $k_1, \ldots, k_n$ such that $\theta_n = k_1 \cdot \phi_1 + \cdots + k_n \cdot \phi_n$ and a number n-tuple $h_1, \ldots, h_n$ such that $\phi_n = h_1 \cdot \theta_1 + \cdots + h_n \cdot \theta_n$.

(iii) If f is in $C_0[a, b]$ and ε is a positive number, then for some positive integer n there exists a number n-tuple $k_1, \ldots, k_n$ such that $\|f - \sum_{p=1}^{n} k_p \cdot \phi_p\|_0 < \varepsilon$.

(iv) If f is in $C_m[a, b]$ and ε is a positive number, there exists a number k such that $\|f - \sum_{p=1}^{n} ((f, \phi_p))_m \cdot \phi_p\|_m < \varepsilon$ if $n > k$.

Definition. Any sequence $\phi_1, \phi_2, \phi_3, \ldots$ of points of $C_m[a, b]$ having properties (i), (iii), and therefore (iv) of the preceding theorem is called a *uniformly complete orthonormal set in $C_m[a, b]$*.

The second expansion theorem of the preceding chapter furnishes another example of a uniformly complete orthonormal set in an inner product space of simple graphs.

Lemma. *Suppose each of f, g, and m is a simple graph with X-projection the interval $[a, b]$, f is continuous, g is continuous, and m is nondecreasing. If h is the simple graph with X-projection $[a, b]$ defined by $h(x) = \int_a^x g \, dm$, $a \leq x \leq b$, then*

$$\int_a^b f \, dh = \int_a^b fg \, dm.$$

Problem. Suppose $\{\phi_p\}_{p=1}^{\infty}$ is a uniformly complete orthonormal set in $C_m[a, b]$ and x is a point of $\mathcal{H}$. Determine a simple graph X in $BV_0[a, b]$ such that for every positive integer p

$$x_p = \int_a^b \phi_p \, dX.$$

Suggestion. Try to guess the solution. We urge the reader to try to settle this problem before looking ahead.

It turns out that there is only one such X. Thus, there exists a reversible and linear transformation from $\mathcal{H}$ to a point set in $BV_0[a, b]$. Moreover, there is an inner product space $H^m[a, b]$, the set of whose points is this point set in $BV_0[a, b]$, that is isometric with $\mathcal{H}$.

The outline given in this chapter is designed to lead the reader to establish this result.

Theorems. *Suppose x is a point of $\mathcal{H}$ and, if n is a positive integer, X_n is the simple graph with X-projection $[a, b]$ defined for each s in $[a, b]$ by*

$$X_n(s) = \sum_{p=1}^{n} x_p \int_a^s \phi_p \, dm.$$

1. X_n is a point in $BV_0[a, b]$.

2. If q is a positive integer and $n \geq q$,

$$x_q = \int_a^b \phi_q \, dX_n.$$

3. *The point sequence in $BV_0[a, b]$, $\{X_n\}_{n=1}^{\infty}$, is a closing sequence in $BV_0[a, b]$. That is, if ε is a positive number, there exists a number k such that $||X_m - X_n||_{bv} < \varepsilon$, if $m > k$ and $n > k$. Hence, $||X_m - X_n||_0 < \varepsilon$ if $m > k$ and $n > k$, so that there exists a simple graph X with X-projection $[a, b]$ such that*

$$X(x) = \sum_{p=1}^{\infty} x_p \int_a^s \phi_p \, dm \quad \text{uniformly for } a \leq s \leq b.$$

4. *X is a point in $BV_0[a, b]$ and if ε is a positive number then there exists a positive integer k such that $||X - X_n||_{bv} < \varepsilon$ if $n > k$.*

5. *If q is a positive integer and $n \geq q$,*

$$\left| x_q - \int_a^b \phi_q \, dA \right| \leq ||\phi_q||_0 \cdot ||X - X_n||_{bv}$$

and therefore $x_q = \int_a^b \phi_q \, dA, q = 1, 2, 3, \ldots$.

6. *If Y is a point of $BV_0[a, b]$ such that $x_q = \int_a^b \phi_q \, dY, q = 1, 2, 3, \ldots$, then if f is in $C_0[a, b]$,*

$$\int_a^b f \, d(X - Y) = 0$$

and so

$$Y = X.$$

If X is a point in $BV_0[a, b]$ and

$$x_p = \int_a^b \phi_p \, dX, \quad p = 1, 2, 3, \ldots,$$

we seek a further condition on X so that $\{x_p\}_{p=1}^{\infty}$ shall be a point x in $\mathcal{H}$, so that there shall exist a number k for which $\sum_{p=1}^{n} x_p^2 \leq k, n = 1, 2, 3, \ldots$.

We write x_p^2 in the form

$$x_p^2 = \int_a^b x_p \phi_p \, dX$$

so that

$$\sum_{p=1}^{n} x_p^2 = \int_a^b \sum_{p=1}^{n} x_p \phi_p \, dX.$$

By the definition of the integral, if ε is a positive number then there exists a finite collection D of nonoverlapping intervals filling up $[a, b]$ such that

$$\int_a^b \sum_{p=1}^n x_p \phi_p \, dX < \varepsilon + \sum \left(\sum_{p=1}^n x_p \phi_p(s) \right) \{X(t) - X(r)\},$$

summed for all the intervals $[r, t]$ in D with $r \leq s \leq t$. Thus,

$$\sum_{p=1}^n x_p^2 < \varepsilon + \sum \left(\sum_{p=1}^n x_p \phi_p(s) \right) Q\{m(t) - m(r)\} \frac{X(t) - X(r)}{Q\{m(t) - m(r)\}}$$

$$< \varepsilon + \left\{ \sum \left(\sum_{p=1}^n x_p \phi_p(s) \right)^2 [m(t) - m(r)] \right\}^{1/2} \left\{ \sum \frac{[X(t) - X(r)]^2}{m(t) - m(r)} \right\}^{1/2}.$$

From this it follows that if there exists a number k such that, for every finite collection D of nonoverlapping intervals filling up $[a, b]$

$$\sum \frac{[X(t) - X(r)]^2}{m(t) - m(r)} \leq k,$$

the sum being taken for all the intervals $[r, t]$ in D, then

$$\sum_{p=1}^n x_p^2 \leq k.$$

We thus have a condition on X that is *sufficient* for x to be a point of $\mathcal{H}$. This condition suggests consideration of an interesting kind of integral.

The Integral $\int_a^b \frac{dX \, dY}{dm}$

Definition. If each of X, Y, and m is a simple graph whose X-projection is the interval $[a, b]$ and if m is increasing, then the integral $\int_a^b \frac{dX \, dY}{dm}$ is a number J such that if ε is a positive number then there exists a finite collection D of nonoverlapping intervals filling up $[a, b]$ such that if D' is a finite collection of nonoverlapping intervals filling up $[a, b]$, with each end of each interval of D an end of some interval of D', then the sum

$$\sum \frac{\{X(t) - X(r)\}\{Y(t) - Y(r)\}}{m(t) - m(r)},$$

formed for all the intervals $[r, t]$ in D', differs from J by less than ε. We shall call this integral *the H-integral of X and Y with respect to m*.

Theorems. *Suppose each of X and m is a simple graph with X-projection the interval $[a, b]$ and m is increasing.*

1. *For the H-integral of X and X with respect to m to exist, it is necessary and sufficient that there should exist a number k such that if D is a finite collection of nonoverlapping intervals filling up $[a, b]$ then*

$$\sum \frac{\{X(t) - X(r)\}^2}{m(t) - m(r)} \leq k,$$

where the sum is taken for all intervals $[r, t]$ in D. If the condition is satisfied, the least such number k is the integral

$$\int_a^b \frac{(dX)^2}{dm}.$$

2. *In order for the H-integral of X and X with respect to m to exist, it is necessary and sufficient that there should exist a simple graph h that is nondecreasing on $[a, b]$ such that if $[r, t]$ is a subinterval of $[a, b]$ then*

$$\{X(t) - X(r)\}^2 \leq \{m(t) - m(r)\}\{h(t) - h(r)\}.$$

3. *If the H-integral of X and X with respect to m exists, then X is continuous at $(t, X(t))$, $a \leq t \leq b$, if m is continuous at $(t, m(t))$.*

4. *If the H-integral of X and X with respect to m exists, then X is of bounded variation on $[a, b]$ and*

$$V_a^b X \leq \{m(b) - m(a)\} \cdot \int_a^b \frac{(dX)^2}{dm}.$$

5. *If the H-integral of X and X with respect to m exists, if the H-integral of Y and Y with respect to m exists and if k is a number, then the H-integrals of X and Y, of $X + Y$ and $X + Y$, and of $k \cdot X$ and $k \cdot X$ with respect to m exist.*

Definition. $H^m[a, b]$ denotes the inner product space in which *point* means an X in $BV_0[a, b]$ such that the H-integral of X and X with respect to m exists (m being the increasing simple graph in $BV_0[a, b]$ previously used) with inner product

$$((X, Y))^m = \int_a^b \frac{dX dY}{dm}.$$

Corollary. *If X is a point of $H^m[a, b]$ and $x_p = \int_a^b \phi_p \, dX$, $p = 1, 2, 3, \ldots$, then $\{x_p\}_{p=1}^\infty$ is a point of $\mathcal{H}$.*

Theorem. *If x is a point in $\mathcal{H}$ and X the point of $BV_0[a, b]$ such that $x_p = \int_a^b \phi_p dX$, $p = 1, 2, 3, \ldots$, then X is in $H^m[a, b]$. Also, if y is a point in $\mathcal{H}$ and Y the point of $H^m[a, b]$ such that $y_p = \int_a^b \phi_p dY$, $p = 1, 2, 3, \ldots$, then*

$$((x, y)) = ((X, Y))^m.$$

That is,

$$\sum_{p=1}^{\infty} \left\{ \int_a^b \phi_p dX \right\} \left\{ \int_a^b \phi_p dY \right\} = \int_a^b \frac{dX dY}{dm}.$$

For each increasing m in $BV_0[a, b]$, the inner product space $H^m[a, b]$ is isometric with $\mathcal{H}$.

15

![black bar]

Mechanical Systems

We consider applications of simple graphs to the analysis of measurable physical things that may vary with time. Each number t is regarded as the measure, in some convenient unit, of the *time* from some specified instant τ, *after* τ if $t > 0$, *before* τ if $t < 0$. Suppose G is a number set each element of which is so regarded. For each t in G, suppose $f(t)$ is the measure (a number) of some physical thing at time t (i.e., at the time from τ determined by t). Then f is a simple graph whose X-projection is G.

Example. Suppose a spherical balloon is being inflated with a gas in such a way that the volume enclosed increases steadily, from a certain instant τ, at the rate of 200 cubic feet per minute. Each number t in the interval $[0, a]$ represents the time measured in minutes from τ. Some simple graphs with X-projection $[0, a]$ that arise are

  i. the volume V such that if t is in $[0, a]$, $V(t)$ is the volume of gas enclosed by the balloon at time t so $V(t) = 200t + V(0)$;

  ii. the surface area s such that if t is in $[0, a]$, $s(t)$ is the area of the surface of the balloon; and

  iii. the radius r such that if t is in $[0, a]$, $r(t)$ is the radius of the balloon.

The simple graphs V, s, and r are related to one another by

$$V = \frac{4}{3}\pi r^3 \quad \text{and} \quad s = 4\pi r^2.$$

175

If f is a simple graph such that for each t in the number set G, $f(t)$ is the measure of a certain physical thing, then $f'(t)$ is the *rate of change* and $f''(t)$ the *acceleration* of this thing at time t. In the example, $V'(t) = 200$, the rate at which gas is being forced into the balloon.

Problems.

1. In the example, suppose c is the time t such that $r(t) = 10$. Find the rate of change and acceleration of the radius r and of the surface area s of the balloon at time c.

2. Suppose that two straight railroad tracks intersect at right angles. At a certain instant, one train is 87 miles from the intersection and approaching it at 40 miles per hour and a second train on the other track is 72 miles from the intersection and approaching it at 30 miles per hour. How fast are the trains approaching one another 2 hours later? What is the minimum distance between them during the entire run? Regard each train as a point and assume that the speed of each is constant.

3. Suppose each of a, b, and h is a positive number. A man is walking along a straight road at the rate of a feet per second. At a certain instant, he is on a bridge h feet directly above a boat that is going at the rate of b feet per second on a straight river at right angles to the road. How fast are the man and boat separating 2 seconds later? Regard the man as a point, the boat as a point, and assume the speed of each is constant.

Falling Bodies

Problems.

1. A body of mass m falls from rest from a place above the earth at a certain instant τ. Suppose s is the simple graph with X-projection the interval $[0, a]$ such that $s(0) = 0$ and if t is a positive number not greater than a, $s(t)$ is the distance that the body falls in time t. Here, a is the least number x such that if $0 < t < x$, the body is above the earth, so that $s(a)$ is the height of the body above the earth at the instant τ. Assuming that the body is drawn toward the earth by the force of gravity and that its motion is retarded by the air resistance that we assume is proportional to the velocity, the simple graph s satisfies the equation

$$ms'' = mg - ks'$$

where each of g and k is a positive number. This is taken for granted. Show that this may be written as

$$\left\{ -\frac{m}{k} L\left[g - \frac{k}{m} s' \right] \right\}' = 1$$

(all simple graphs being restricted to $[0, a]$). There is a number c such that

$$-\frac{m}{k} L\left[g - \frac{k}{m} s' \right] = I + c,$$

$$c = -\frac{m}{k} L(g),$$

and

$$s' = \frac{mg}{k} \left\{ 1 - E\left[-\frac{k}{m} I \right] \right\}.$$

Hence show that if t is in $[0, a]$,

$$s(t) = \frac{mg}{k} \left\{ t - \frac{m}{k} + \frac{m}{k} e^{-\frac{m}{k} t} \right\}.$$

2. Show that if air resistance is neglected,

$$s(t) = \frac{1}{2} g t^2, \quad 0 \le t \le a.$$

3. Consider the same problem under the hypothesis that the air resistance is proportional to the square of the velocity, so that the equation for s is

$$ms'' = mg - k(s')^2, \quad k > 0.$$

Denote the number $\frac{k}{mg}$ by c^2 and show that this equation may be written as

$$\frac{s''}{2} \left\{ \frac{1}{1 - cs'} + \frac{1}{1 + cs'} \right\} = g$$

or

$$\left\{ \frac{1}{2c} L\left[\frac{1 - cs'}{1 + cs'} \right] \right\}' = -g,$$

and that if t is in $[0, a]$,

$$s(t) = \frac{m}{k} L\left\{ C\left(\sqrt{\frac{kg}{m}} t \right) \right\}$$

or, as it is frequently written,

$$s(t) = \frac{m}{k} \log \cosh \sqrt{\frac{kg}{m}} t.$$

4. A body of mass m is suspended by a spring and the system is in equilibrium. At a certain instant τ, the body is pulled downward a distance h feet and released. Suppose s is the simple graph whose X-projection is the set of nonnegative numbers such that $s(0) = h$, $s'(0) = 0$, and, if t is a positive number, $s(t) = 0$, $s(t) = x$, or $s(t) = -x$ according as the body is at the equilibrium position, below the equilibrium position the distance x, or above the equilibrium position the distance x, respectively, at time t after the instant τ. Assume that if air resistance and friction are neglected, $s'' + k^2 s = \underline{0}$, where k is a positive number, and show that if t is a nonnegative number, $s(t) = h \cdot C(kt)$, or, as is frequently written,

$$s(t) = h \cos kt.$$

5. If, in the preceding problem, account is taken of the resistance of the medium in which the system is suspended, the equation for s is $s'' + r^2 s' + k^2 s = \underline{0}$, where r is a positive number. Find s if $k^2 > \frac{r^2}{4}$.

6. Suppose that the system is subject to a continuous impressed force. Then the equation to be solved is $s'' + r^2 s' + k^2 s = g$, where g is a continuous simple graph whose X-projection is the set of nonnegative numbers. Determine s.

Fluid Pressure

A region R in a vertical plane is submerged in water. Suppose a is the distance from the water level to the upper boundary of R and b is the distance from the water level to the lower boundary of R. See Figure 15.1.

We assume that

i. if $[p, q]$ is a subinterval of $[a, b]$, the *pressure* on the strip S, that is the subset of R whose upper and lower boundaries are horizontal and at depths p and q, respectively, is the weight of a column of water of cross section S and height some number between p and q, and

ii. if $S_1, \ldots, S_n$ is any finite collection of nonoverlapping strips of this kind filling up R and P_i is the pressure on S_i, $i = 1, \ldots, n$, then the pressure on R is $P_1 + \cdots + P_n$.

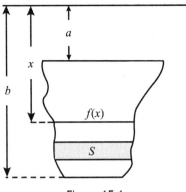

Figure 15.1

Problems.

1. Suppose f is the simple graph with X-projection $[a, b]$ such that for each x in $[a, b]$, $f(x)$ is the width of R measured horizontally x units below the surface of the water and $g(x) = \int_a^x f$. Let W denote the density of water. Show that the pressure on R is

$$W \cdot \int_a^b I \, dg.$$

2. A trough of trapezoidal cross-section is 2 feet deep, 2 feet wide at the bottom, and 3 feet wide at the top. If the trough is full of water, what is the pressure on the end of the trough?

3. A cylindrical tank 8 feet in diameter is lying on its side. If it contains water to a depth of 6 feet, find the pressure of the water on one end of the tank.

Motion in a Plane

Suppose $[0, a]$ is an interval and if t is in $[0, a]$, $(x(t), y(t))$ is a point representing the location of a particle in a plane at time t, measured from some instant τ. See Figure 15.2. Each of x and y is a simple graph with X-projection $[0, a]$. The point set to which P belongs only if there is a number t in $[0, a]$ such that $P = (x(t), y(t))$ is the *path* of the motion of the particle and is denoted by $\{x, y\}$. The ordered pair $\{x', y'\}$ is the *velocity* and the ordered pair $\{x'', y''\}$ is the *acceleration* of the motion. If t is in $[0, a]$, the number $Q\{[x'(t)]^2 + [y'(t)]^2\}$ is the *speed* or *magnitude of the velocity* at time t

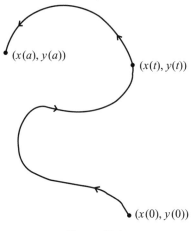

Figure 15.2

and the number $Q\{[x''(t)]^2 + [y''(t)]^2\}$ is the *magnitude of the acceleration* at time t.

The statement that the path $\{x, y\}$ has *length* means that there exists a number k such that if D is a finite collection of nonoverlapping intervals filling up $[0, a]$, the sum $\sum Q\{[x(q) - x(p)]^2 + [y(q) - y(p)]^2\}$, formed for all the intervals $[p, q]$ in D, does not exceed k. If $\{x, y\}$ has length and $[u, v]$ is a subinterval of $[0, a]$, the length of $\{x, y\}$ on $[u, v]$, denoted by $\ell_u^v\{x, y\}$, is the least number k such that the above sum formed for any finite collection of nonoverlapping intervals filling up $[u, v]$ does not exceed k. If $\{x, y\}$ has length on $[0, a]$ and c is a number between 0 and a, then

$$\ell_0^a\{x, y\} = \ell_0^c\{x, y\} + \ell_c^a\{x, y\}.$$

If each of x' and y' has X-projection $[0, a]$ and is continuous, then

$$\ell_0^t\{x, y\} = \int_0^t Q[(x')^2 + (y')^2], \quad 0 \le t \le a.$$

If we denote this by $s(t)$, then

$$s' = Q[(x')^2 + (y')^2],$$

so that $s'(t)$ *is the speed of the motion at time* t.

Hypothesis. We suppose x' and y' have X-projection $[0, a]$ and are continuous, and $\{x'(t)\}^2 + \{y'(t)\}^2 > 0$ for every t in $[0, a]$.

Problems.

1. Show that there exists only one continuous simple graph θ with X-projection $[0, a]$ such that $0 \leq \theta(0) < 2\pi$ and for each number t in $[0, a]$

$$x'(t) = s'(t) \cdot \cos \theta(t)$$

and

$$y'(t) = s'(t) \cdot \sin \theta(t).$$

(That is, $x' = s'C[\theta]$ and $y' = s' S[\theta]$.)

2. Suppose t_1 and t_2 are numbers in $[0, a]$, $P_1 = (x(t_1), y(t_1))$, $P_2 = (x(t_2), y(t_2))$, $\overline{P_1 P_2} = Q\{[x(t_1) - x(t_2)]^2 + [y(t_1) - y(t_2)]^2\}$, the distance from P_1 to P_2, and $\widehat{P_1 P_2} = |\int_{t_1}^{t_2} Q[(x')^2 + (y')^2]|$, the length of the path between P_1 and P_2. Show that if c is a positive number and t_1 is in $[0, a]$, there exists a positive number d such that if t_2 is in $[0, a]$, is distinct from t_1, and differs from t_1 by less than d, then the quotient

$$\frac{\overline{P_1 P_2}}{\widehat{P_1 P_2}}$$

differs from 1 by less than c.

3. Show that if c is a positive number and $0 \leq t_1 < a$, then there exists a positive number d such that if $t_1 < t_2 \leq a$ and $t_2 - t_1 < d$, then

$$\left| \frac{x(t_2) - x(t_1)}{\overline{P_1 P_2}} - \cos \theta(t_1) \right| < c$$

and

$$\left| \frac{y(t_2) - y(t_1)}{\overline{P_1 P_2}} - \sin \theta(t_1) \right| < c.$$

Thus, the direction of the motion at time t_1 is along the tangent line to the path at the point $(x(t_1), y(t_1))$. See Figure 15.3.

4. Suppose t is a number in the X-projection of x'' and in the X-projection of y''. Show that

(i) $\theta'(t) = \dfrac{x'(t)y''(t) - x''(t)y'(t)}{\{s'(t)\}^2}$,

(ii) $s''(t) = x''(t) \cdot \cos \theta(t) + y''(t) \cdot \sin \theta(t)$,

and

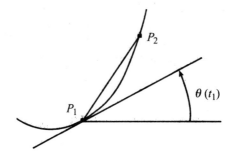

Figure 15.3

(iii) $s'(t)\theta'(t) = -x''(t) \cdot \sin \theta(t) + y''(t) \cdot \cos \theta(t).$

If

$$\alpha_m(t) = s''(t) = x''(t) \cdot \cos \theta(t) + y''(t) \cdot \sin \theta(t)$$

and

$$\alpha_n(t) = s'(t) \cdot \theta'(t) = -x''(t) \cdot \sin \theta(t) + y''(t) \cdot \cos \theta(t),$$

we see that the magnitude of the vector $\{\alpha_m(t), \alpha_n(t)\}$ is the magnitude of the acceleration $\{x''(t), y''(t)\}$. The component of the acceleration normal (perpendicular) to the path is $\alpha_n(t)$ and the component tangential to the path is $\alpha_m(t)$. See Figure 15.4.

The *curvature* of the path $\{x, y\}$ at the point $(x(t), y(t))$ is

$$\kappa(t) = D_u \theta[\sigma],$$

where σ is the simple graph to which the point (p, q) belongs only if (q, p) belongs to s and $\sigma(u) = t$; i.e., $u = s(t)$.

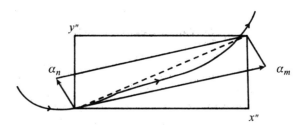

Figure 15.4

Problems.

1. Show that

$$\kappa(t) = \frac{x'(t)y''(t) - x''(t)y'(t)}{\{s'(t)\}^3}.$$

2. Show that the normal component of acceleration is

$$\alpha_n(t) = \kappa(t) \cdot \{s'(t)\}^2.$$

3. Investigate the motion with path $\{x, y\}$ if $x(0) = 0$, $y(0) = 0$, $x'(0) = v$, $y'(0) = w$, $x'' = \underline{0}$, and $y'' = -g$, where each of v, w, and g is a positive number.

4. Investigate the motion of a point on the rim of a wheel of radius a feet rolling along a straight road making b revolutions per second.

The curvature of a straight line at each of its points is 0. The curvature of a circle at each of its points is the reciprocal of its radius.

Centroids

The statement that s is the *line interval* with ends P_1 and P_2 means that P_1 and P_2 are points and s is the point set to which P belongs only if there exists a number t in the interval $[0, 1]$ such that $P = (1 - t) \cdot P_1 + t \cdot P_2$. The statement that the point set M is *convex* means that if P_1 and P_2 are points of M, the line interval with ends P_1 and P_2 is a subset of M.

Examples. The point set containing only the point P is convex; the set of all points is convex; a line interval is convex; if M_1 and M_2 are convex point sets having a common part M, then M is convex.

Definition. *The particle of mass m at the point P is the ordered pair* (P, m) where P is a point and m is a nonnegative number.

The statement that (P, m) is the *centroid* of the finite set $(P_1, m_1), \ldots,$ (P_n, m_n) of particles means that $m = m_1 + \cdots + m_n > 0$ and

$$P = \frac{m_1 \cdot P_1 + \cdots + m_n \cdot P_n}{m_1 + \cdots + m_n}.$$

The point P belongs to the line interval with ends P_1 and P_2 only if there is a particle at P_1 and a particle at P_2 such that their centroid is at P.

Problems.

1. Suppose r is a positive integer and n is a positive integer greater than r, $(P_1, m_1), \ldots, (P_r, m_r)$ has centroid (P', m'), and $(P_{r+1}, m_{r+1}), \ldots,$ (P_n, m_n) has centroid (P'', m''). Show that the centroid of $(P_1, m_1), \ldots,$ (P_n, m_n) is the centroid of the two particles (P', m') and (P'', m'').

2. Suppose n is a positive integer, $P_1, \ldots, P_n$ are n points, and M is the point set to which P belongs only if P is at the centroid of n particles at $P_1, \ldots, P_n$. Then M is convex and any convex set that contains $P_1, \ldots, P_n$ includes M as subset.

3. Extend the preceding considerations to points in space.

Definition. Suppose R is a bounded point set and, for each point P in R, $(P, \rho(P))$ is a particle at P. The centroid of this set of particles is called the centroid of R with respect to the density function ρ and is the particle (S, m) such that

$$m = \iint\limits_R \rho \ dJ \quad \text{and} \quad S = \frac{\iint\limits_R P\rho \ dJ}{\iint\limits_R \rho \ dJ}$$

where $\iint\limits_R P\rho \ dJ$ is the point (u, v) such that if

$$h(x, y) = x \cdot \rho(x, y)$$

and

$$k(x, y) = y \cdot \rho(x, y),$$

then

$$u = \iint\limits_R h \ dJ \quad \text{and} \quad v = \iint\limits_R k \ dJ.$$

Problems.

1. Let R denote the region $[Q; 0, 1]$ determined by the simple graph Q and the interval $[0, 1]$ and let ρ denote the density function defined by $\rho(x, y) = x$. Find the centroid of R with respect to ρ.

2. Find the centroid of the circular disc defined by $x^2 + y^2 \leq 1$ with respect to the density function $\rho(x, y) = 1$.

3. Find the centroid of the quarter-circular disc defined by $x^2 + y^2 \leq 1, x \geq 0, y \geq 0$, with respect to the density function $\rho(x, y) = 1$.

4. Suppose G is a finite collection of nonoverlapping rectangular intervals filling up the rectangular interval $[ab; cd]$ and the region R is a subset of $[ab; cd]$. Suppose $R_1, \ldots, R_n$ are the subsets of R included in rectangular intervals of G, and (S_i, m_i) is the centroid of R_i with respect to the density function ρ. Show that the centroid of R with respect to ρ is the centroid of the finite set of particles $(S_1, m_1), \ldots, (S_n, m_n)$. See Figure 15.5.

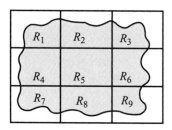

Figure 15.5

Integral Tables

The statement that F is an *antiderivative* of the simple graph f means that F is a simple graph such that $F' = f$. If the X-projection of f is an interval and F is an antiderivative of f, then $F + \underline{c}$, where c is a number, is an antiderivative of f and there are no others.

A table of indefinite integrals is a table of antiderivatives. We denote an antiderivative of the simple graph f by the symbol

$$\int f.$$

Here are some antiderivatives followed by the corresponding entries in one of several available tables.

1. $\displaystyle\int \underline{a} = aI$ $\displaystyle\int a\,dx = ax$

7. $\displaystyle\int I^n = \frac{I^{n+1}}{n+1}$ $\displaystyle\int x^n\,dx = \frac{x^{n+1}}{n+1}$ (except $n = -1$)

9. $\displaystyle\int \frac{1}{I} = L[|I|]$ $\displaystyle\int \frac{dx}{x} = \log x \text{ or } \log(-x)$

11. $\displaystyle\int E = E$ $\displaystyle\int e^x = e^x$

12. $\displaystyle\int L = IL - I$ $\displaystyle\int \log x\,dx = x\log x - x$

16. $\displaystyle\int \frac{1}{a^2 + I^2} = \frac{1}{a}A\left[\frac{I}{a}\right]$ $\displaystyle\int \frac{dx}{a^2 + x^2} = \frac{1}{a}\tan^{-1}\frac{x}{a}$

183. $\displaystyle\int T = -L[|C|] = L[|K|]$ $\displaystyle\int \tan x\,dx = -\log\cos x = \log\sec x$

187. $\displaystyle\int S^2 = \frac{1}{2}(I - CS)$ $\displaystyle\int \sin^2 x\,dx = -\frac{1}{2}\sin x \cos x + \frac{x}{2}$

202. $\displaystyle\int \frac{1}{1+C} = T\left[\frac{I}{2}\right]$ $\displaystyle\int \frac{dx}{1+\cos x} = \tan\frac{x}{2}$

244. $\displaystyle\int T^2 = T - I$ $\displaystyle\int \tan^2 x\, dx = \tan x - x$

313. $\displaystyle\int S = C$ $\displaystyle\int \sinh x\, dx = \cosh x$

In using the tables, it is necessary to supply the range on which the formulas are valid. For instance, if $n = \frac{1}{2}$ in the antiderivative of I^n, the formula is valid for all nonnegative numbers but not for negative numbers.

Index of Simple Graphs

Unless it is otherwise stated, the following letters denote certain specific simple graphs:

$H, L, E, Q, I, \mathcal{S}, \mathcal{C}, \mathcal{T}, \mathcal{K}, \mathcal{A}, \Omega, A, T, K, C, S, B$, as follows.

Symbol	Meaning	Page
H	(x, y) belongs to H only if x is a positive number and $y = \dfrac{1}{x}$	12
L	(x, y) belongs to L only if x is a positive number and $y = \displaystyle\int_1^x H$	15
E	(x, y) belongs to E only if (y, x) belongs to L	19
Q	(x, y) belongs to Q only if x is nonnegative number and $y = \sqrt{x}$	21
I	(x, y) belongs to I only if x is a number and $y = x$	30
$\mathcal{S}$	$\mathcal{S} = \frac{1}{2}\left\{E - \frac{1}{E}\right\}$	40
$\mathcal{C}$	$\mathcal{C} = \frac{1}{2}\left\{E + \frac{1}{E}\right\}$	40
$\mathcal{T}$	$\mathcal{T} = \frac{\mathcal{S}}{\mathcal{C}}$	40
$\mathcal{K}$	$\mathcal{K} = \frac{1}{\mathcal{C}}$	40
$\mathcal{A}$	(x, y) belongs to $\mathcal{A}$ only if (y, x) belongs to $\mathcal{T}$	40
Ω	$\Omega = \dfrac{1}{1 + I^2}$	43

189

Symbol	Meaning	Page
A	(x, y) belongs to A only if x is a number and $$A(x) = \int_0^x \Omega$$	43
T	(x, y) belongs to T only if $-\frac{\pi}{2} < x < \frac{\pi}{2}$ and (y, x) belongs to A or there is a number t between $-\frac{\pi}{2}$ and $\frac{\pi}{2}$ and an integer n such that $x = t + n\pi$ and $A(x) = A(t)$	46
K	$K^2 = 1 + T^2$, $K(x) > 0$ if $-\frac{\pi}{2} < x < \frac{\pi}{2}$, $K(x) < 0$ if $\frac{\pi}{2} < x < \frac{3\pi}{2}$ and if there is a number t distinct from $\frac{\pi}{2}$ between $-\frac{\pi}{2}$ and $\frac{3\pi}{2}$ and an integer n such that $x = t + 2n\pi$, then $K(x) = K(t)$	49
C	If x is the abscissa of a point of K, then $C(x) = \frac{1}{K(x)}$ and, otherwise, $C(x) = 0$	49
S	If x is the abscissa of a point of T, $S(x) = C(x)T(x)$ and, if n is an integer, $S\left\{\frac{(2n-1)\pi}{2}\right\} = (-1)^{n-1}$.	50
B	(x, y) belongs to B only if y is in the interval $\left[-\frac{\pi}{2}, \frac{\pi}{2}\right]$ ans $x = S(y)$	50

Glossary of Definitions

The statement that:	Means:
(x, y) *is a point*	(x, y) is an ordered number pair
x is the *abscissa* of the point P	x is the first or left-most number of P
y is the *ordinate* of the point P	y is the second or right-most number of P
S is a *point set*	S is a collection of one or more points
f is a *simple graph*	f is a point set no two points of which have the same abscissa
$f(x)$ is f of x	$f(x)$ is the ordinate of that point of the simple graph f whose abscissa is x
M is the *X-projection* of the point set S	M is the number set to which x belongs only if x is the abscissa of a point of S
$[a, b]$ is an *interval*	a and b are numbers, $a < b$, and $[a, b]$ is the number set to which x belongs only if x is a, x is b, or x is a number between a and b
$[f; a, b]$ is the *region* determined by the simple graph f and the interval $[a, b]$	the X-projection of f includes $[a, b]$ and $[f; a, b]$ is the point set to which (x, y) belongs only if x is in $[a, b]$ and y is 0, y is $f(x)$, or y is a number between 0 and $f(x)$
$\underline{h}$, read h *horizontal*, is a horizontal line	h is a number and $\underline{h}$ is the point set to which (x, y) belongs only if x is a number and y is h
x is an *end* of the interval $[a, b]$	x is a or x is b

G is a collection of non overlapping intervals	G is a collection of one or more intervals and, if two intervals in G have a number in common, this number is an end of each of them
S is a *finite* set	there is a positive integer n such that S does not contain n elements
S is an *infinite* set	if n is a positive integer, S contains n elements
S is an *inner sum* for the region $[H; a, b]$	there exists a finite collection G of nonoverlapping intervals filling up $[a, b]$ such that if the length of each interval $[p, q]$ in G is multiplied by $\frac{1}{q}$, the sum of all the products so formed is S
$\int_a^b H$ is the *area* of the region $[H; a, b]$	$\int_a^b H$ is the least number that no inner sum for $[H; a, b]$ exceeds
g is the straight line of slope m containing the point (a, b)	m is a number and g is the simple graph to which (x, y) belongs only if x is a number and $y = m(x - a) + b$
$\log_a x$ is the *logarithm of x to the base a*	a is a positive number distinct from 1, x is a positive number, and $$\log_a x = \frac{L(x)}{L(a)}$$
a is the number e	a is the positive number such that $L(a) = 1$
a^x is *a to the x*	x is a number and $a^x = E\{x L(a)\}$
$|x|$ is the *absolute* value of x	x is a number and $|x| = Q(x^2)$
$h|$, read *h vertical*, is a vertical line	h is a number and $h|$ is the point set to which (x, y) belongs only if y is a number and x is h
the simple graph f has *slope* at the point P	P is a point of f such that each two vertical lines with P between them have between them a point of f distinct from P and there exists a number m, called the *slope* of f at P, such that if α is a straight line containing P of slope greater than m and B is a straight line containing P of slope less than m, then there exist two vertical lines $h|$ and $k|$ with P between them such that every point of f between $h|$ and $k|$ distinct from P is between α and β

g is the *tangent line* to the simple graph f at the point P	f has slope m at P and g is the straight line of slope m containing P
$f + g$ is the *sum* of the simple graph f and the simple graph g	the X-projection of f and the X-projection of g have a common part and, if x is in this common part, $(f + g)(x) = f(x) + g(x)$
$f \cdot g$ or fg is the *product* of the simple graph f and the simple graph g	the X-projection of f and the X-projection of g have a common part and, if x is in this common part, $(f \cdot g)(x) = f(x)g(x)$
$f[g]$ is the *bracket product* f of g	there is a number x such that $g(x)$ is in the X-projection of f and, if x is such a number, $f[g](x) = f\{g(x)\}$
$\frac{1}{g}$ is the *reciprocal* of the simple graph g	there is a number x such that $g(x) \neq 0$ and, if x is such a number, $\frac{1}{g}(x) = \frac{1}{g(x)}$
$\frac{f}{g}$ is the *quotient* of the simple graph f by the simple graph g (or over g)	$\frac{f}{g} = f \cdot \frac{1}{g}$
f' is the *derivative* of the simple graph f	there is a number x such that f has slope at $(x, f(x))$ and, for any such x, $f'(x)$ is the slope of f at $(x, f(x))$
s is the *segment* with ends a and b	a and b are numbers and s is the number set to which x belongs only if x is between a and b
the simple graph f has property S at P (or is *continuous* at P)	P is a point of f such that if $\underline{\alpha}$ and $\underline{\beta}$ are horizontal lines with P between them, there exist vertical lines $h\|$ and $k\|$ with P between them such that every point of f between $h\|$ and $k\|$ is between $\underline{\alpha}$ and $\underline{\beta}$
$g\|_a^b$ is the *g-length* of the interval $[a, b]$	g is a simple graph whose X-projection includes $[a, b]$ and $g\|_a^b = g(b) - g(a)$
g is *nondecreasing* on $[a, b]$	the X-projection of the simple graph g includes $[a, b]$ and the g-length of every subinterval of $[a, b]$ is nonnegative
g is *increasing* on $[a, b]$	the X-projection of the simple graph g includes $[a, b]$ and the g-length of every subinterval of $[a, b]$ is positive

F is a *transformation* from the set A to the set B	F is a collection of ordered pairs (a, b) whose leftmost element a is in A, whose rightmost element b is in B, such that each element of A is the first element of *only one* ordered pair in F and each element of B is the second element of *some* ordered pair in F
f is *g-integrable* on $[a, b]$	each of f and g is a simple graph whose X-projection includes the interval $[a, b]$ and there exists a number J, called the integral from a to b of f with respect to g and denoted by $\int_a^b f \, dg$, such that if c is a positive number, there exists a finite collection D of nonoverlapping intervals filling up $[a, b]$ such that if D' is a finite collection of nonoverlapping intervals filling up $[a, b]$ with each end of each interval of D an end of some interval of D' and the g-length of each interval of D' is multiplied by the ordinate of any point of f whose abscissa is in that interval, then the sum of all the products so formed differs from J by less than c
f is a *simple surface*	f is a transformation from a point set to a number set
$[ab; cd]$ is a *rectangular interval*	$[a, b]$ is an interval, $[c, d]$ is an interval, and $[ab; cd]$ is the point set to which (x, y) belongs only if x belongs to $[a, b]$ and y to $[c, d]$
R is the *edge* of the rectangular interval $[ab; cd]$	R is the point set to which (x, y) belongs only if (x, y) belongs to $[ab; cd]$ and $x = a$, $x = b$, $y = c$, or $y = d$
s is a *rectangular segment.*	s is a rectangular interval minus its edge
the simple surface f is *continuous* at $(P, f(P))$	if c is a positive number, there exists a rectangular segment s containing P such that if Q is a point in s and in the XY-projection of f, then $f(P)$ differs from $f(Q)$ by less than c
f_1' is the *1-derivative* of the simple surface f	f_1' is the simple surface to which $((x, y), z)$ belongs only if $z = D_x f[I, y]$
f_2' is the *2-derivative* of the simple surface f	f_2' is the simple surface to which $((x, y), z)$ belongs only if $z = D_y f[x, I]$

f_{ij}'' is the *ij-derivative* of the simple surface f	$f_{ij}'' = (f_i')_j'$ $(i = 1, 2,\ j = 1, 2)$
the simple surface f has *gradient* at $(P, f(P))$, $P = (x, y)$	there exists only one ordered number pair $\{p, q\}$ [called the gradient of f at $(P,\ f(P))$] such that if c is a positive number, there exists a rectangular segment s containing P such that if (u, v) is a point of s in the XY-projection of f, then $$f(u, v) - f(x, y) = p \cdot (u - x) + q \cdot (v - y)$$ $$+ |P - (u, v)| \cdot \left[\begin{array}{c}\text{a number between} \\ -c \text{ and } c\end{array}\right]$$
$g\vert_a^b\vert_c^d$ is the *g-area* of the rectangular interval $[ab; cd]$	g is a simple surface whose XY-projection includes $[ab; cd]$ and $g\vert_a^b\vert_c^d = \{g[b, I] - g[a, I]\}\vert_c^d = g(b, d) - g(a, d) - g(b, c) + g(a, c)$